Seismic Hazard of the Circum-Pannonian Region

Edited by
Giuliano F. Panza
Mircea Radulian
Cezar-Ioan Trifu

2000

Birkhäuser Verlag
Basel · Boston · Berlin

Reprint from Pure and Applied Geophysics
(PAGEOPH), Volume 157 (2000), No. 1/2

Editors:

Giuliano F. Panza
Department of Earth Sciences
Via Weiss, 4
34127 Trieste
Italy

Mircea Radulian
National Institute for Earth Physic
Box MG – 2
76900 Bucharest
Romania

Cezar-Ioan Trifu
ESG Canada Inc.
1 Hyperion Ct.
K7K 7G3 Kingston, ON
Canada

A CIP catalogue record for this book is available from the Library of Congress, Washington
D.C., USA

Deutsche Bibliothek Cataloging-in-Publication Data

Seismic hazard of the Circum-Pannonian region / ed. by.Giuliano F. Panza ... - Basel ;
Boston ; Berlin : Birkhäuser, 2000
   (Pageoph topical volumes)
   ISBN 3-7643-6263-4 (Basel ...)
   ISBN 0-8176-6263-4 (Boston)

© 2000 Birkhäuser Verlag, P.O.Box 133, CH-4010 Basel, Switzerland
Printed on acid-free paper produced from chlorine-free pulp. TCF ∞
Printed in Germany

9 8 7 6 5 4 3 2 1

Contents

Pure appl. geophys. 157 (2000) 1–4
0033–4553/00/020001–04 $ 1.50 + 0.20/0

Preface

The Circum-Pannonian Region extends over a relatively large territory from Central to Eastern Europe including densely populated and industrialized areas in several countries. While the highest seismic hazard is controlled at a regional scale by the large Vrancea intermediate-depth earthquakes, several shallow earthquake-prone areas are very important at a local scale. In the present geopolitical context of Europe, the harmonization of Western and Eastern Europe in terms of seismic safety compliance appears crucial for the future development and prosperity of the whole continent. This special volume summarizes the outcome of several international projects conducted over recent years, such as *Quantitative seismic zoning of the Circum-Pannonian Region* (EC-Copernicus), *Earthquake hazard associated with the Vrancea region seismicity* (NATO), and *Microzonation of Bucharest, Russe and Varna cities in connection with Vrancea Earthquakes* (NATO).

The above projects have resulted in a high degree of innovation. Effective state-of-the-art techniques have been developed for the assessment of seismic hazard, and reliable ground motion estimates were obtained. This collection gathers fourteen original studies which offer quantitative information required for the design, construction and retrofitting of the built environment that will greatly reduce the number of human casualties and the amount of property loss due to a large earthquake that may occur in this region or its vicinity. In particular, it is important to outline the impact of these studies on the reduction of the environmental hazard associated with the existing four nuclear power plants in the region. As such, the results obtained should be considered a starting point for subsequent and more detailed investigations into the retrofitting of the nuclear plants in Bulgaria, Hungary, Romania and Slovenia. Additionally, these studies have significantly contributed to the establishment of the source and response spectra to be used in connection with the large intermediate-depth earthquakes generated by the Vrancea region of Romania. The results also suggest the working hypotheses that could be further employed for an integration and revision of the European Building Code EC8.

An initial group of papers employs the analysis of the seismicity pattern, main geologic structures and geodynamic models, in their attempt to characterize the seismogenically active areas and define the provinces of seismogenic homogeneity. As such, the introductory study by *Meletti et al.* outlines a working methodology that has been employed in the seismotectonic zonation of Italy for hazard assessment purposes. The paper by *Poljak et al.* integrates information on seismicity, earthquake mechanisms, displacement rates and stress estimates to delineate and characterize the seismogenic areas of the territory of Slovenia. Similarly, *Radulian et al.* review and update the available information in their effort to identify the principal features of the seismogenic areas active on the territory of Romania, with a special reference to the Vrancea zone.

The next group of papers includes a few theoretical observational studies aimed to providing additional support for the delineation of earthquake-prone areas, source parameters, and structural models that may be further used in seismic hazard assessment. The work by *Gorshkov et al.* represents an application of the morphostructural analysis to the block-structure of the regional crust. Their results appear to correlate well with the recorded seismicity, except for the northeastern zone of the Vrancea region, where significantly lower seismic intensities have been recorded to date. The dynamics of the Vrancea region is the subject of a study by *Soloviev et al.* Using a block model, they concluded that a variation in model parameters has little effect on the orientation of the fracture slip for the intermediate-depth events. Based on a 2-D finite-element model of a sinking slab, *Ismail-Zadeh et al.* find a good correlation between the depth distribution of stress in Vrancea and the recorded seismicity and energy release. However, since the annual cumulative seismic moment estimate far exceeds that expected for a pure phase-transition, they consequently suggest dehydration of rocks as the triggering mechanism of these events. On a different topic, *Živčić et al.* use surface-wave dispersion analysis to derive the velocity distribution beneath the territory of Slovenia.

The main focus of this volume is the quantitative seismic zoning. Traditional methods use either a deterministic or probabilistic approach, based on empirically derived laws for ground motion attenuation. The work by *Musson* is a good exemplification of the probabilistic approach, the results of which are subsequently compared and found in good agreement, except for the Vrancea zone, with the results of a deterministic approach. In the case of Vrancea, the attenuation relations used in the probabilistic approach seem to underestimate, mainly at large distances, the seismic hazard due to the intermediate-depth earthquakes, whereas the deterministic results seem representative of the most conservative scenario. Recent advances in computer technology, however, now make possible the use of the deterministic numerical synthesis of ground motion for seismic hazard calculations. The deterministic approach capably addresses aspects largely overlooked in the probabilistic approach, such as: (a) the effect of crustal properties on attenuation: (b) the derivation of ground motion parameters from synthetic time histories, instead of using highly simplified attenuation functions; (c) the direct evaluation of resulting maps in terms of design parameters, without requiring the adaptation of probabilistic maps to design ground motions; and (d) the generalization of design parameters to locations where there is little seismic history. Maximum displacements, velocities, and, based on the European Building Code EC8, design ground acceleration maps have thus been produced by *Živčić et al.* for Slovenia, *Markušić et al.* for Croatia, *Bus et al.* for Hungary, and *Radulian et al.* for Romania.

The last two contributions in the volume are dedicated to studies of local site effects that could affect the microzonation of large urban areas. *Moldoveanu et al.* employed a technique based on the modal summation and finite differences to calculate the expected ground motion in the capital city of Bucharest due to large intermediate-depth Vrancea earthquakes. Their results outline that the presence of alluvial sediments and the possible variation of event scenario require the use of all three components of motion for a reliable determination of the seismic input. The study of *Marmureanu et al.*, more limited in scope, offers a laboratory analysis of the attenuation effects for surface layers. The authors confirm that seismic attenuation in sedimentary layers is a function of the strain levels induced by large earthquakes, and find that the quality factor is nearly constant over a relatively wide frequency range, between 7 and 100 Hz.

Acknowledgements

This collection of papers gathers contributions from numerous authors, who are acknowledged both for the quality of their work and the interest in contributing to a special volume. Additionally, this volume was made possible due to the dedicated work of numerous reviewers whose time, conscientious efforts and scientific judgement have been oriented to ensuring that the scientific community will circulate a sound and efficient message. As such, *Karim Aoudia*, *Kuvvet Atakan*, *Gail Atkinson*, *Igor Beresnev*, *Guenter Bock*, *Jean Bonnin*, *David Boore*, *Hilmar Bungum*, *Peter Byrne*, *Armando Cisternas*, *Rodolfo Console*, *Torsten Dahm*, *Catherine Dorbath*, *John Ebel*, *Mariana Eneva*, *Bob Engdhal*, *Mustafa Erdik*, *Andrew Feustel*, *Liam Finn*, *Clifford Frohlich*, *Gotfried Grunthal*, *Susan Hough*, *Steven Jaume*, *Andrzej Kijko*, *Raul Madariaga*, *Tina Nunziata*, *David Oppenheimer*, *Robert Pearce*, *David Perkins*, *Avi Shapira*, *Hong Kie Thio*, *Colin Thomson*, *Augustin Udias*, *Jean Virieux*, *Friedemann Wenzel*, *Jiri Zahradnick* are warmly thanked for willingly accepting the above responsibility and providing competent reviews. Finally, we would like to express our gratitude to *Miguel Angel Virasoro*, Director of the Abdus Salam International Center for Theoretical Physics (ICTP), to *Giuseppe Furlan*, Head of the ICTP program for Training and Research in Italian Laboratories, and to *Lucio Delcaro*, Rector of the University of Trieste, for their encouragement and support.

Giuliano F. Panza
Department of Earth Sciences
Via Weiss, 4
I-34127 Trieste
Italy
E-mail: panza@geosun0.univ.trieste.it

M. Radulian
National Institute for Earth Physic
Box MG-2
76900 Bucharest
Romania
E-mail: mircea@infp.infp.ro

Cezar Trifu
ESG Canada Inc.
1 Hyperion Ct.
K7K 7G3 Kingston, ON
Canada
E-mail: trifu@esg.ca

 To access this journal online:
http://www.birkhauser.ch

Pure appl. geophys. 157 (2000) 5-9
0033-4553/00/020005-05 $ 1.50 + 0.20/0

| Pure and Applied Geophysics

Introduction

The complex logistic problem connected with transnational seismic zoning and microzoning following standardized criteria has greatly benefited from the existing organizational network established in the framework of the Earth Sciences Committee of the Central European Initiative.

The results contained in this special issue offer information necessary to greatly reduce the number of human casualties and the amount of property loss upon the occurrence of a big earthquake in a large part of Southeast Europe and North-Africa.

Given the number of nuclear power plants located in the studied region, the results of the present study *should* be used as a starting point for successive more detailed investigations aimed at the retrofitting of the existing plants. This may be a necessary action in order to reduce the environmental hazard associated with such plants.

Maps of various seismic hazard parameters numerically modelled, and whenever possible tested against observations, such as peak ground displacement, velocity and acceleration, of practical use for the design of earthquake-safe structures, have been produced, in combination with the first microzoning actions in large cities, such as Bucharest, Ljubljana and Sofia. The *Realistic Modelling of Seismic Input for Megacities and Large Urban Areas is presently a major commitment of UNESCO-IGCP, under its project 414.*

The synoptic analysis of the seismicity pattern, of the main geologic structures and of the geodynamic models, provided the starting point for the characterization of the seismogenically active areas and the means to define the provinces of seismogenic homogeneity. A regional seismic catalogue has been compiled with national catalogues, and earthquake mechanism and size have been determined for each seismogenic area, with the key contribution of local experts. The simultaneous involvement of scientists from the different countries has allowed a minimization of the effects of political boundaries, quite often hampering such studies.

In Figures 1–3 we show maps of the peak values of horizontal motion (displacement, D, velocity V, and design ground acceleration, DGA) for the European/Mediterranean countries that have contributed to this major effort for the mitigation of seismic hazard. For more details, see the national studies.

The peak values of D, V and DGA, and pertinent periods T(D) and T(V), at the sites where nuclear power plants are located are summarized in Figure 4. The values obtained at Cernavoda, Kozloduy and Paks are controlled by the intermediate-depth Vrancea events ($M = 7.7$)

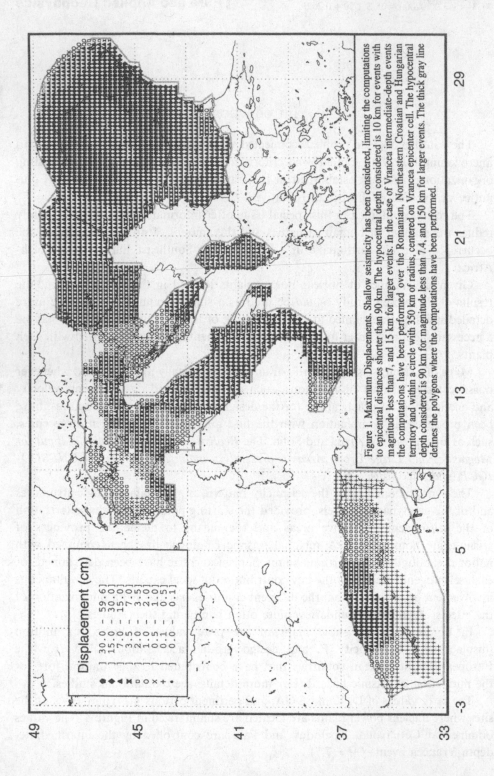

Figure 1. Maximum Displacements. Shallow seismicity has been considered, limiting the computations to epicentral distances shorter than 90 km. The hypocentral depth considered is 10 km for events with magnitude less than 7, and 15 km for larger events. In the case of Vrancea intermediate-depth events the computations have been performed over the Romanian, Northeastern Croatian and Hungarian territory and within a circle with 350 km of radius, centered on Vrancea epicenter cell. The hypocentral depth considered is 90 km for magnitude less than 7.4, and 150 km for larger events. The thick gray line defines the polygons where the computations have been performed.

Displacement (cm)

●	30.0 – 59.6
◆	15.0 – 30.0
◀	7.0 – 15.0
⋈	3.5 – 7.0
□	2.0 – 3.5
○	1.0 – 2.0
×	0.5 – 1.0
+	0.1 – 0.5
·	0.0 – 0.1

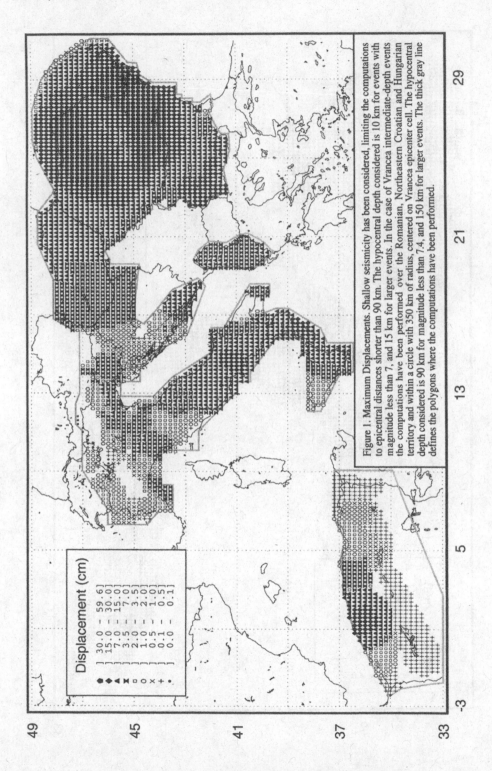

Figure 1. Maximum Displacements. Shallow seismicity has been considered, limiting the computations to epicentral distances shorter than 90 km. The hypocentral depth considered is 10 km for events with magnitude less than 7, and 15 km for larger events. In the case of Vrancea intermediate-depth events the computations have been performed over the Romanian, Northeastern Croatian and Hungarian territory and within a circle with 350 km of radius, centered on Vrancea epicenter cell. The hypocentral depth considered is 90 km for magnitude less than 7.4, and 150 km for larger events. The thick gray line defines the polygons where the computations have been performed.

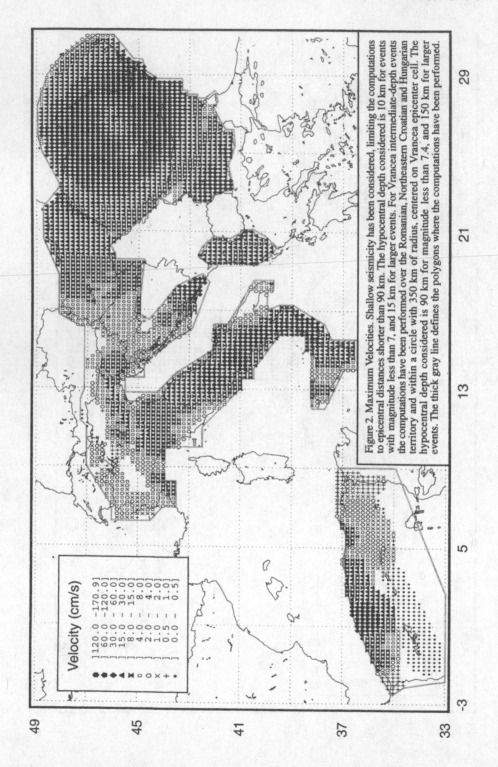

Figure 2. Maximum Velocities. Shallow seismicity has been considered, limiting the computations to epicentral distances shorter than 90 km. The hypocentral depth considered is 10 km for events with magnitude less than 7, and 15 km for larger events. For Vrancea intermediate-depth events the computations have been performed over the Romanian, Northeastern Croatian and Hungarian territory and within a circle with 350 km of radius, centered on Vrancea epicenter cell. The hypocentral depth considered is 90 km for magnitude less than 7.4, and 150 km for larger events. The thick gray line defines the polygons where the computations have been performed.

Velocity (cm/s)

120.0	–	170.9
60.0	–	120.0
30.0	–	60.0
15.0	–	30.0
8.0	–	15.0
4.0	–	8.0
2.0	–	4.0
0.5	–	2.0
0.0	–	1.0
0.0	–	0.5

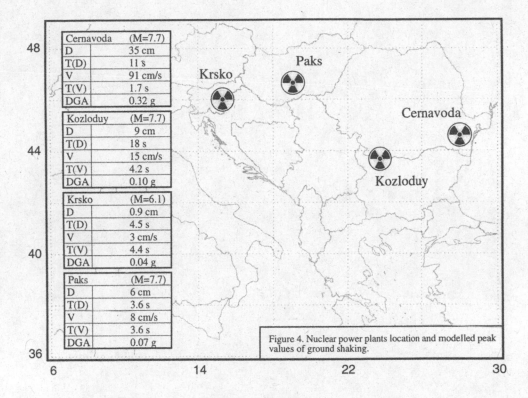

Cernavoda	(M=7.7)
D	35 cm
T(D)	11 s
V	91 cm/s
T(V)	1.7 s
DGA	0.32 g

Kozloduy	(M=7.7)
D	9 cm
T(D)	18 s
V	15 cm/s
T(V)	4.2 s
DGA	0.10 g

Krsko	(M=6.1)
D	0.9 cm
T(D)	4.5 s
V	3 cm/s
T(V)	4.4 s
DGA	0.04 g

Paks	(M=7.7)
D	6 cm
T(D)	3.6 s
V	8 cm/s
T(V)	3.6 s
DGA	0.07 g

Figure 4. Nuclear power plants location and modelled peak values of ground shaking.

Acknowledgements

The comprehensive maps defining the deterministic seismic hazard have been also made possible by the contribution of Prof. B. Muço (for Albania, unpublished). Special thanks are extended to Acad. Prof. V. I. Keilis-Borok who attracted our attention to the problem of the seismic hazard connected with the Vrancea earthquakes.

G. F. Panza
Dipartimento di Scienze della Terra
Università di Trieste
and
The Abdus Salam International Center for Theoretical Physics

F. Vaccari
Dipartimento di Scienze della Terra
Università di Trieste
and
Gruppo Nazionale per la Difesa dai Terremoti
CNR-Rome

Pure appl. geophys. 157 (2000) 11–35
0033–4553/00/020011–25 $ 1.50 + 0.20/0

Pure and Applied Geophysics

Construction of a Seismotectonic Model: The Case of Italy

CARLO MELETTI,[1] ETTA PATACCA[2] and PAOLO SCANDONE[2]

Abstract—Procedures for constructing a seismotectonic model of Italy, designed to be used as a basis for hazard assessment, are described. The seismotectonic analysis has essentially been based on a GIS-aided cross-correlation of three data sets concerning:
– the 3-D structural model of Italy and surrounding areas;
– the space distribution of historical and present seismicity;
– the kinematic model of the Central Mediterranean region, referred to the last 6 Ma and including the available information on the present-day plate motion and stress field.
The seismicity pattern in the study area is controlled by a quite complex geodynamic framework which includes:
– continent–continent convergence (Alps and Dinarides) with development of a neutral arc bordering the plate margins;
– plate divergence across margins characterized by passive slab sinking (Northern Apennines and Calabrian Arc), with development of backarc basins (Northern Tyrrhenian Sea and Southern Tyrrhenian Sea) flanked by forelandward migrating thrust belt-foredeep systems;
– plate divergence across a margin previously characterized by lithosphere sinking and afterwards discharged from the subducted slab (Southern Apennines), with development of quite peculiar rift processes within the inactive thrust belt;
– transpression (Northern Sicily) due to the combined effect of plate convergence (Africa-Europe) and high-rate flexure-hinge retreat of an intervening plate (Adria microplate) with high angles between the respective slip vectors;
– intraplate strain partition and fault activity (mainly combined strike-slip and thrust motions), possibly in correspondence of inverted structures.
The results of the seismotectonic analysis are synthesized in a zonation of Italy in which every delimited zone corresponds to the surface projection of a kinematically-homogeneous segment of a seismogenic fault system. In Cornell-type hazard evaluations every polygon should be considered as a homogeneous source-zone, seat of randomly-distributed earthquakes. A homogeneous mechanical behaviour of an entire zone and a random earthquake-distribution within a single source zone obviously represent oversimplified assumptions since every zone includes one or more master-fault segments responsible for the greatest events in the area and several second-order associated faults responsible for the background minor seismicity. Therefore, major faults and background seismicity should be treated separately. Nevertheless, the oversimplified assumption of homogeneous seismic zones was the price the authors consciously paid to produce, in a reasonably short time, a homogeneous product relative to the entire national territory, suitable for earthquake hazard evaluation and for decisions regarding risk mitigatiton.

Key words: Seismotectonics, Italy, kinematic model, seismic zonation, seismic hazard.

[1] CNR, Gruppo Nazionale per la Difesa dai Terremoti, Via Nizza, 128-Roma, Italy.
[2] Dipartimento di Scienze della Terra, Via S. Maria, 53-Pisa, Italy.

1. Foreword

The seismotectonic zonation of an earthquake-prone region, that is the identification, delimitation and characterization of seismogenic geological structures, is a preliminary basic operation for achieving realistic assessments of the seismic hazard. Obviously, the adopted seismic zonation influences, and in some cases (see CORNELL, 1968; BENDER and PERKINS, 1987) closely controls, the distribution pattern of the hazard results. Unfortunately, neither a univocal approach nor standard procedure has been established to date by the scientific community (BASHAM and GIARDINI, 1993; GIARDINI and BASHAM, 1993) in order to produce in different countries seismotectonic zonations based on homogeneous or at least comparable criteria (see, e.g., GRELLET *et al.*, 1993; GRÜNTHAL *et al.*, 1995; LENHARDT, 1995; PAPAZACHOS, 1988, 1996; SCANDONE *et al.*, 1992; VOGT and GODEFROY, 1981). Consequently, different parameters have been preferred by different researchers in order to perform seismotectonic investigations. In some cases this is due to the wide variety of geological scenarios (geodynamic regime, tectonic style, kinematic evolution, etc.), but more often it is because of dissimilar basic assumptions and philosophical approaches (see, among many others, ALLEN, 1976; BORISOV *et al.*, 1976; BUNE *et al.*, 1974; CISTERNAS *et al.*, 1985; D'OFFIZI, 1994; GELFAND *et al.*, 1972; HAYS, 1980; IAEA SAFETY GUIDE, 1991; LAVECCHIA *et al.*, 1994; MUIR-WOOD, 1993; MULARGIA *et al.*, 1987; PANTOSTI and YEATS, 1993; PATACCA and SCANDONE, 1986; SCHICK, 1978; SCHWARTZ and COPPERSMITH, 1986; SLEMMONS and DEPOLO, 1986; TRIFONOV and MACHETTE, 1993; WALSH and WATTERSON, 1988; WELLS and COPPERSMITH, 1994; WORKING GROUP ON CALIFORNIA EARTHQUAKE PROBABILITIES, 1995; ZOBACK, 1993).

In this paper we shall briefly describe procedures and results of seismotectonic analysis of Italy carried out within the framework of the scientific activities of the CNR-Gruppo Nazionale per la Difesa dai Terremoti (National Group for the Defense against Earthquakes, National Council of Research). The results of this analysis are synthesized in a zonation of Italy (Figs. 8, 9) in which every delimited zone corresponds to the surface projection of a kinematically-homogeneous segment of a seismogenic fault system (see SCANDONE *et al.*, 1992). In Cornell-type hazard evaluations every polygon should be considered as a homogeneous source-zone, seat of randomly-distributed earthquakes. A homogeneous mechanical behaviour of an entire zone and a random earthquake-distribution within a single source zone obviously represent oversimplified assumptions since every zone includes one or more master-fault segments responsible for the greatest events in the area, and several second-order associated faults responsible for the background minor seismicity. Therefore, major faults and background seismicity should be treated separately. Actually, current research has been addressed to discriminate between capable fault segments and associated minor faults, as well as to investigate the kinematic and mechanical behaviour of the active master faults responsible for the major ($M \geq 5.5$) crustal earthquakes. Nevertheless, the available paleoseismological information (see among others BRUNAMONTE *et al.*, 1991; CELLO *et al.*, 1997; CINTI *et al.*, 1997; FERRELI *et al.*, 1996; GALADINI *et al.*, 1996; GHISETTI, 1992; MICHETTI *et al.*, 1996; PANTOSTI *et al.*, 1993, 1996; SERVA *et al.*, 1986; VALENSISE and PANTOSTI, 1992; WARD and VALENSISE, 1989) is still too patchy for the application of hybrid methods of hazard evaluation (e.g., PERUZZA *et al.*, 1997) over the entire Italian territory. Therefore, the oversimplified assumption of homogeneous seismic zones was the price we consciously paid to produce, in a reasonable short time, a homogeneous product relative to the entire national territory. In the present state of the art, the zonation of Figures 8 and 9 has been adopted as an input framework for earthquake hazard evaluations (COSTA *et al.*, 1993; PANZA *et al.*, 1998; ROMEO and PUGLIESE, 1997; SLEJKO, 1996; SLEJKO *et al.*, 1998) used by political authorities for decisions regarding risk mitigation in Italy. In addition, seismotectonic regionalization combined with earthquake epicentres and focal mechanisms has been used for intermediate-term earthquake forecasting (BOSCHI *et al.*, 1995b; COSTA *et al.*, 1995, 1996).

The aim of this paper is to discuss the methodology we followed for the seismotectonic analysis and to describe the general results in terms of source geometry and kinematics. Special emphasis will be placed on the contribution given to the seismotectonic analysis by a kinematic approach (PATACCA and SCANDONE, 1986) in order to fix reliable constraints to the possible correlation between seismicity and active geological structures in complex thrust-and-fold belts like the Alps and the Apennines.

2. Geological Framework

The Neogene-Quaternary kinematic evolution and the present-day stress field of the Central Mediterranean area, together with the tectonic structure of Italy (Fig. 1), have often been described as a direct result of the Africa-Europe convergence (e.g., BEN AVRAHAM *et al.*, 1990; BOCCALETTI and DAINELLI, 1982; DEWEY *et al.*, 1973, 1989; MAZZOLI and HELMAN, 1994). Nevertheless, the available geological and geophysical information suggests a more complex plate interaction, with an important role played by the intervening Adria microplate (see ANDERSON and JACKSON, 1987a). In addition, other geodynamic processes, mostly related to a passive sinking of the subducting Adria lithosphere, must be taken into account. In the Apennines these processes are responsible for the migration of the thrust belt/foredeep system towards the Padan-Adriatic-Ionian foreland, and for the synchronous opening of the Tyrrhenian backarc basin according to slip vectors largely exceeding the values of the Africa-Europe convergence (see, among many others, MALINVERNO and RYAN, 1986; MANTOVANI *et al.*, 1996; PATACCA and SCANDONE, 1986, 1989; PATACCA *et al.*, 1990, 1993). The present-day seismicity in

the Central Mediterranean region (AMATO *et al.*, 1997; ING, 1995) mostly follows
the principal mountain chains, that is the Alps, the Apennines and the Dinarides,
though clusters of epicentres indicate a certain fragmentation of the foreland areas,
notably in Southern Sicily and in the Sicily Channel, in the Gargano-Tremiti region
and in the Central Adriatic Sea (Fig. 2).

The Alps are a well-known thrust-and-fold belt comprising a huge pile of
basement and cover nappes transported towards the European foreland, detached
from the lower (Europe) and the upper (Adria) plates, as well as from the
Jurassic-Cretaceous Tethys Ocean (see CNR, P.F. GEODINAMICA, 1990). Recent

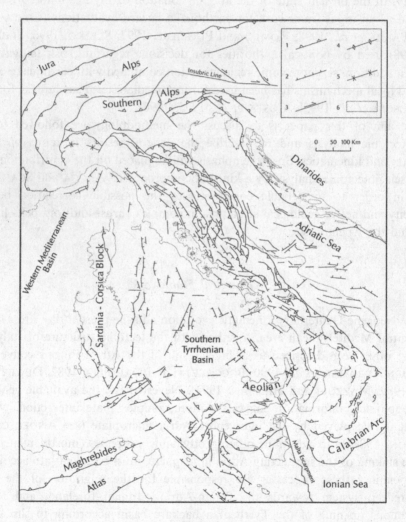

Figure 1

Structural sketch of Italy and surrounding regions. 1 thrusts; 2 normal faults; 3 strike-slip faults; 4
Pliocene-Quaternary anticlines; 5 Pliocene-Quaternary synclines; 6 backarc basins floored by oceanic
crust.

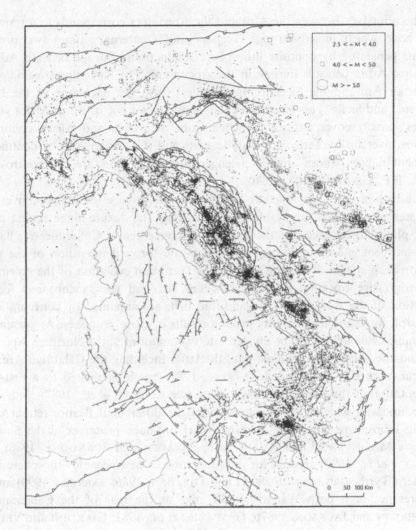

Figure 2
Present-day seismicity in Italy and surrounding areas plotted on the structural sketch of Figure 1.
Seismicity from ING (1995).

reflection seismic profiles across the Western and Central Alps (ROURE *et al.*, 1990;
SCHMID *et al.*, 1996) display a clear image of the plate boundary in this sector, with
a deep-seated triangle zone responsible for the back-thrusting of the Adria lower
crust in the Ivrea zone and for the piling up of south-verging imbricate fans in the
Southern Alps originated by detachment processes in the upper plate. As regards
the Southern Alps, the maximum shortening has been calculated in the eastern part
of the system (CASTELLARIN, 1978) where the Alpine structures join those of the
Dinarides, and where the strongest historical earthquakes of the Alps have been
recorded (e.g., 1976 Friuli earthquake, $M = 6.5$).

The boundary between the Alps and the Apennines corresponds to a transform fault zone which linked through Tertiary and Quaternary times two orogenic systems generated by opposite lithosphere subductions: Europe beneath Adria in the Alps; Adria beneath Europe in the Apennines. The Apennines (including the Calabrian Arc and the Sicilian Maghrebides) form the backbone of the Italian Peninsula and Sicily. This mountain chain constitutes a pile of Adria/Africa verging nappes, mostly cover nappes, detached from the Adria and Africa continental margins, overlain by Tethyan ophiolitic units and (Calabrian Arc) by continental-basement nappes whose original paleogeographic domains are still controversial (CNR, P.F. GEODINAMICA, 1990).

Subduction of Adria beneath Europe started because of active convergence processes in a neutral arc system, but with time the flexure-hinge retreat of the lower plate largely exceeded the amount of convergence. Consequently, backarc extension took place, accompanied by high-rate forward migration of the thrust belt-foredeep system. Referring to the post-Tortonian evolution of the Apennines, the progressive opening of the Tyrrhenian Sea and the synchronous forward migration of the Apennine thrust-and-fold belt, accompanied by consumption of the Adria foreland, are the most striking results of these processes. At present, the Apenninic chain appears to be divided into two major arcs: the Northern Apenninic Arc and the Southern Apenninic Arc, the latter including the Calabrian Arc. This configuration (PATACCA and SCANDONE, 1989) has been related to a first-order segmentation of the subducting lower plate (see ROYDEN *et al.*, 1987), with major free boundaries which have accommodated the differential flexure retreat of the sinking lithosphere. High-rate (≥ 5 cm/year) roll-back processes in the Southern Arc (see MALINVERNO and RYAN, 1986; PATACCA and SCANDONE, 1986, 1989; PATACCA *et al.*, 1990) account for the generation of new oceanic lithosphere in the Southern Tyrrhenian basin (FINETTI and DEL BEN, 1986; SARTORI, 1990) and for the presence of a deep Wadati-Benioff zone at the rear of the Calabrian Arc (ANDERSON and JACKSON, 1987b; GASPARINI *et al.*, 1982; GIARDINI and VELONÀ, 1991).

The northern, northeastern and southwestern margins of the present Adria microplate are well defined by the Western Alps-Southern Alps, Dinarides and Apennines, respectively. The southern continuation of the microplate, on the contrary, is still a matter of debate. Geodetic and stress field investigations in the Central Medidterranean region (MONTONE *et al.*, 1997; MÜLLER *et al.*, 1992; RAGG *et al.*, 1995; WARD, 1994) suggest that Southern Sicily belongs to the Africa plate and is moving NW with respect to Europe. Conversely, the same types of data indicate that Apulia is moving NE at high angles with the slip vectors describing the Africa motion (see ANZIDEI *et al.*, 1997; WARD, 1994). Several rotation poles of Adria versus Europe and of Africa versus Europe are available in the geological literature (Tables 1 and 2). In this paper we propose a new rotation pole of Adria versus Europe, located some tens of kilometres SW of Genoa, which in our opinion

Table 1

Rotation poles of Adria versus Europe

Reference	Lat.	Long.
ANDERSON and JACKSON (1987a, b)	45.8	10.2
WESTAWAY (1990)	44.5	9.5
WARD (1994)	46.8	6.3
This work	44.2	8.3

better fits the bulk of the geological information as well as the available VLBI data from Matera station (WARD, 1994) and the fault plane solutions of major earthquakes along the outer margin of Adria from the Western Alps to the Western Hellenides (DZIEWONSKI *et al.*, 1983 and subsequent fault plane solutions available in the CMT computer file catalogue; EVA *et al.*, 1997; HERAK *et al.*, 1995; MUCO, 1994; PAPADIMITRIOU, 1993; RENNER and SLEJKO, 1994). According to this reconstruction, the Ionian Sea is part of the present-day Adria microplate whose sinking slab is shown by the South-Tyrrhenian Wadati-Benioff zone (ANDERSON and JACKSON, 1987b; GASPARINI *et al.*, 1982; GIARDINI and VELONÀ, 1991) and by the Aeolian calc-alkaline volcanic arc (SERRI, 1997). The Malta Escarpment, reaching in the north the active Etna volcano (HIRN *et al.*, 1997) and forming the most striking morphotectonic feature of the region (SCANDONE *et al.*, 1981), appears to be the best candidate for transtensional plate margin between Africa (Southern Sicily) and Adria (Ionian Sea).

The Sicilian Maghrebides (LENTINI *et al.*, 1996) represent the natural westward continuation of the Apennines. Nevertheless, their structure and kinematic evolution are quite different from the Apennines because the Sicilian orogenic segment has evolved, starting from Late Tortonian-Messinian times, as a transpressional

Table 2

Rotation poles of Africa versus Europe

Reference	Lat.	Long.
MCKENZIE (1972)	22.7	−28.2
DEWEY *et al.* (1973)	31.3	−34.7
MINSTER and JORDAN (1978)	25.5	−21.2
CHASE, 1978	29.2	−23.5
SEARLE (1980)	21.3	−21.0
LIVERMORE and SMITH (1985)	22.7	−31.9
ANDERSON (1985)	27.6	−19.7
HELMAN (1989)	0.6	−15.8
DE METS *et al.* (1990)	21.0	20.6
WESTAWAY (1990)	21.0	−21.0
WARD (1994)	−7.7	−57.1

belt accreted between the northwestward moving Africa margin (Africa-Europe convergence) and the eastward-escaping Calabrian Arc (roll-back of the subducting Ionian lithosphere).

3. Base Data and Geological/Geophysical Constraints for Seismotectonic Analysis

The seismotectonic analysis of the Italian territory is basically founded on GIS-aided cross correlation of three data sets concerning:
– the 3-D structural model of Italy and surrounding areas;
– the space distribution of historical and present seismicity;
– the kinematic model of the Central Mediterranean region, referred to the last 6 million years and including the available information on the present-day plate motion and stress field.

Figure 3 is a flow chart explaining the research activities which led to the construction of the seismotectonic model of Italy.

The structural model allows us to recognise and define lateral inhomogeneities both in the crust and in the mantle, to establish the geometry of the potentially active structures and to evaluate the mechanical characteristics of the rocks at

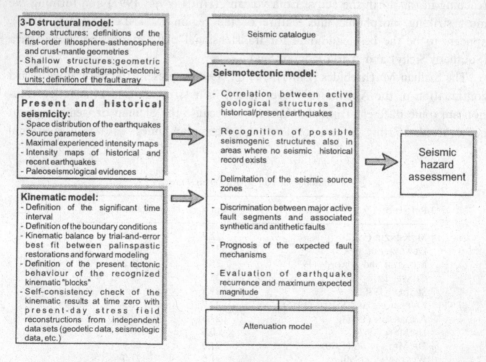

Figure 3
Flow chart of the research activities which led to the seismotectonic modelling of Italy.

different depths. Owing to the different investigation techniques, the model resolution is obviously different going from the lithosphere/asthenosphere boundary to the upper crust (see CHIARABBA and AMATO, 1996; CNR, P.F. GEODINAMICA, 1990; NICOLICH, 1989; SUHADOLC and PANZA, 1989). As regards shallow depths (≤ 10–15 km), the information has been improved by commercial seismic lines and exploratory wells.

The basic documents we used to analyse the earthquake distribution and to correlate seismicity and active geological structures are principally represented by the ING Catalog (ING, 1995) as regards the present-day seismicity, the CNR-Progetto Finalizzato Geodinamica Catalog (POSTPISCHL, 1985) which covers the 1000–1980 time interval and the NT Catalog (CAMASSI and STUCCHI, 1996), a new declustered catalog currently adopted for hazard evaluations in Italy. Other important pieces of information are represented by a map of the maximal experienced intensities in Italy (MOLIN et al., 1996) and by the intensity maps of several hundred historical and recent earthquakes which have struck the territory over the last 2000 years (MONACHESI and STUCCHI, 1997; BOSCHI et al., 1995a, 1997). The macroseismic field reconstructions were very useful for constraining the seismotectonic model and for delimiting the seismic source zones. In several cases close correlations between active geological structures and earthquakes were found; in other cases the existence of an accurate macroseismic documentation forced us to better explore areas where no active faults had been previously recognized. In some cases, finally, the absence of earthquake documentation in areas characterized by severe recent deformation suggested the need for improvements in the historical research, as well as new plans for accurate paleoseismological investigations.

A kinematic model based on reliable palinspastic restorations provides the time/space trend of several independent and dependent variables (e.g., rate of flexure-hinge retreat of the subducting lithosphere, migration rate of the compressional and extensional fronts, tectonic subsidence and uplift, slip rates along active faults, etc.) which may be very important for seismotectonic investigations. The availability of well-calibrated curves representing the time-space variation of these parameters, in fact, allows us to better control the extreme points of the functions at time zero, that is, it enables us to better understand the present-day tectonic activity and, in addition, to recognize possible changes of the geodynamic regime. Moreover, a reliable kinematic model may establish first-order boundary conditions that force us to reconsider unquestioned postulates and sometime to abandon them as out of data opinions. The most striking example concerning the last point is represented by the Apenninic "paradox." From Eastern Liguria to the Calabrian Arc the Apennines border the southwestern margin of the subducting Adria plate. Consequently, the Adria margin in this region has usually been described as a converging plate margin. However, if we consider the north and northeastern margins of the same microplate, we see that it is also bordered by compressional features which point to another converging plate boundary extended from the

Western Alps to the Western Hellenides. The geometry and kinematic evolution of the Circum-Adriatic thrust-and-fold belts, on the other hand, cannot be explained as a result of the crashing of the Adria microplate between the Africa and Europe major plates. The results of the kinematic analysis, in addition, demonstrate that Adria has undergone counterclockwise rotation at least since Neogene times, and this motion implies extensional and not compressional slip vectors across the southwestern margin of the microplate. If this reconstruction is correct, how and why did a contractional belt take place in the Apennines while synchronous thrust-and-fold belts developed in the Southern Alps and Dinarides? The kinematic paradox is only apparent if we take into account that the tectonic style changed across the inner margin of Adria from a neutral arc system to an arc/backarc system. This change implies that the rate of the flexure-hinge retreat in the lower (Adria) plate largely exceeded the convergence rate and caused the opening of a backarc basin. The relationships among the principal kinematic parameters are expressed by:

$$V_{ext} = V_{flr} - V_c$$

where V_{ext} is the extension rate in the backarc basin (higher than 5 cm/year in the Southern Tyrrhenian Sea), V_{flr} is the velocity of flexure-hinge retreat in the lower plate and V_c is the Europe-Adria plate convergence rate. The rate of migration of the thrust belt-foredeep system towards the foreland areas roughly equals the rate of the flexure hinge retreat of the lower plate. In a regime of passive slab sinking (which is expected in west-dipping subductions, see DOGLIONI, 1991), flexure-hinge retreat may also continue under negative values of the convergence rate; the only difference in the kinematic balance is represented by the fact that the increase of source area absorbed by the overall extension in the backarc basin is transferred as sink area at the outer margin of the diverging plate. In conclusion, the kinematic analysis shows that the post-Tortonian Apenninic compression did not take place along a converging margin but along a diverging one, while the Southern Alps and Dinarides developed along a converging margin. In our opinion, the Apenninic "paradox" may be justified not only in terms of kinematics, but also in terms of mechanics by the geometry and dynamics of the lithosphere-asthenosphere system. The deep structure of the mountain chain, in fact, shows that the shallow asthenosphere occupying the space between the upper plate and the decoupled lower plate (MARSON *et al.*, 1995; SUHADOLC *et al.*, 1993) may act as a pushing back-stop of the contractional system driven by the passive lithosphere sinking (MELETTI *et al.*, 1995). Figure 4 shows a schematic lithospheric section across the Northern Apenninic Arc in which the edge of the deep-seated asthenospheric wedge plays the role of the leading edge of the shallow thrust system.

It is interesting to note that the kinematic analysis suggests a persistence of the described geodynamic regime in the Northern Apenninic Arc and probably in the Calabrian Arc, while it exhibits a dramatic change in the Southern Apennines

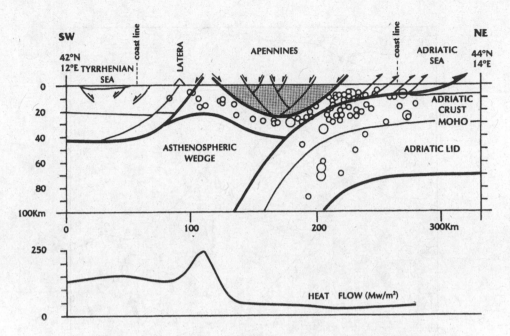

Figure 4
Schematic lithospheric section across the Northern Apenninic Arc (after MELETTI *et al.*, 1995).

(CINQUE *et al.*, 1993; HIPPOLYTE *et al.*, 1994) which took place near the base of the Middle Pleistocene. In this region a model similar to the one in the Northern Apennines has recently been proposed by DOGLIONI *et al.* (1996). We disagree as regards this model because the available geological information indicates that the flexure-hinge retreat suddenly ceased around 0.65 Ma (slab detachment?) from Eastern Abruzzi to Southern Basilicata and a generalized uplift of the mountain chain followed, accompanied by a regional tilting of the whole edifice towards the NE and by normal faulting along the Tyrrhenian slope. At present, active faults roughly follow the orographic divide of the mountain chain in an extensional stress field characterized by NE-SW oriented *T*-axes (AMATO and MONTONE, 1997; MONTONE *et al.*, 1997). We have interpreted (PATACCA *et al.*, 1997) this extensional stress field as the consequence of an early stage of rifting, related to the counterclockwise rotation of Adria, which began to happen approximately 0.65 Ma after the subduction processes stopped in the Southern Apennine segment.

4. *Seismotectonic Model and Seismic Zonation*

Figure 5 represents a structural/kinematic sketch of the Italian peninsula and surrounding areas which shows the first-order tectonic structures. The following major elements have been indicated:

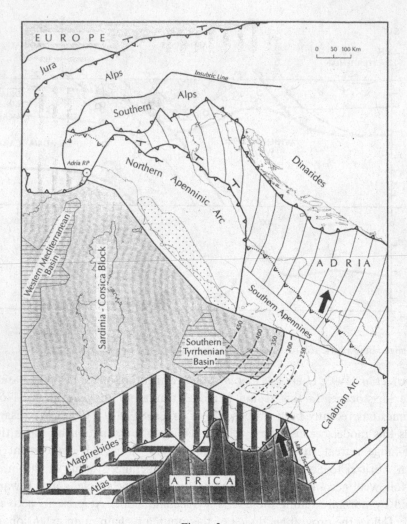

Figure 5

Structural/kinematic sketch of Italy and surrounding areas showing the traces of the slip vectors of Africa versus Europe (according to LIVERMORE and SMITH, 1985) and Adria versus Europe (according to the rotation pole proposed in this paper). See explanation in the text.

– the Adria microplate and the traces of the slip vectors describing its motion towards Europe;
– the rotation pole of Adria (Adria RP);
– the northern portion of the Africa plate and the traces of the slip vectors describing the Africa-Europe convergence. Black arrows represent the slip vectors at Matera (Adria) and Noto (Africa) stations according to WARD (1994);
– the European plate, including the Corsica-Sardinia block as well as the Western Mediterranean and the Tyrrhenian backarc basins;

- the Malta Escarpment, interpreted as the plate boundary between Africa and Adria;
- the thrust-and-fold belts, together with the major lithospheric free boundaries;
- the shallow asthenospheric wedges (dotted areas) in the Northern Apenninic Arc and the Calabrian Arc;
- the compression front of the Europe-verging Alpine system;
- the compression fronts of the Adria-verging outer thrust systems (Southern Alps, Dinarides);
- the Insubric Line, roughly separating the Europe-verging Alpine nappes from the Adria-verging thrust sheets of the Southern Alps;
- the compression fronts of the Adria-verging inner thrust system (Northern Apenninic Arc and Calabrian arc). The compression front is inactive in the Southern Apennines;
- the extensional young fault system between the Adria and Europe plates in the Southern Apennines;
- the Wadati-Benioff zone of the Southern Tyrrhenian Sea;
- the compression front of the Africa-verging Maghrebides.

Figures 6 and 7 show the present-day and historical earthquakes plotted on the geological features of Figure 5.

In the proposed seismotectonic model, the earthquakes bordering the outer margin of Adria are attributed to thrusts and to transpressional faults, all related to the counterclockwise rotation of Adria versus Europe. Due to the position of the rotation pole, the slip vectors obviously increase from the Western Alps to the Dinarides under the same angular velocity. As regards Italian territory, maximum shortening and maximum seismic potential are expected north of the Venice Gulf, in accordance with present and historical seismicity.

Conversely, no plate convergence may be invoked in order to explain the observed seismicity in the Apennines. In the Northern Apenninic Arc, the slab sinking with a flexure-hinge retreat faster than the Adria divergence may wholly justify the regional seismicity pattern characterized by:

- low/medium-energy compressional earthquakes along the Padan-Adriatic margin of the Apennines related to active frontal and lateral ramps which branch off from the sole-thrust at greater and greater depths (but in any case not exceeding about 20 kilometers) moving from the foreland towards the mountain chain;
- medium/high-energy earthquakes, mostly displaying extensional dip-slip focal mechanisms, in correspondence to an axial belt located between the Adria flexure-hinge and the Tyrrhenian asthenospheric wedge. The bulk of the focuses is contained in a crustal synform where the opposite geometries of the rising Tyrrhenian asthenosphere and the sinking Adria lithosphere are accommodated (see Figs. 1 and 2) by NE-dipping low angle master-faults and SW-dipping high-angle antithetical-faults (BARCHI et al., 1996; BONCIO et al., 1996; MELETTI et al., 1995);

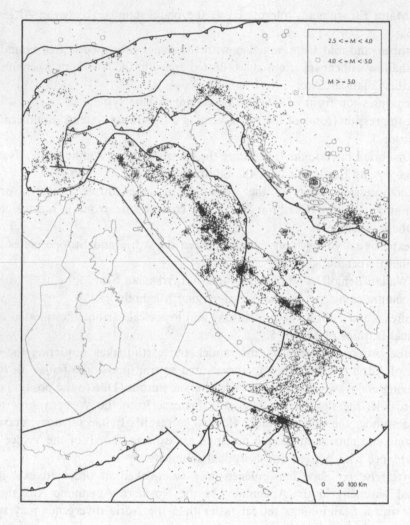

Figure 6
Present-day seismicity plotted on the sketch of Figure 5. Seismicity from ING (1995).

low-energy very shallow earthquakes above and behind the mobile asthenospheric wedge.

In the Southern Apennines, the cessation of subduction while the counterclockwise rotation of the Adria microplate was still continuing produced a strong modification of the lithosphere-asthenosphere system and the establishment of an extensional regime. A seismic axial belt is present, characterized by medium/high-energy earthquakes whose available focal solutions show dip-slip mechanisms. We relate the seismicity of the Southern Apennines to very young normal faults generated by the Adria divergence, superimposed on inactive contractional structures.

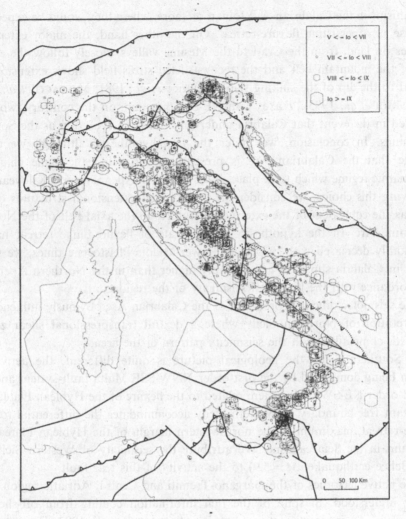

Figure 7
Historical seismicity plotted on the sketch of Figure 5. Seismicity from POSTPISCHL (1985).

The present-day behaviour of the Calabrian Arc is a matter of debate. Due to conflicting geological/geophysical evidence, it is not clear whether the well-known Wadati-Benioff zone of the Southern Tyrrhenian Sea is still attached to the Ionian lithosphere or if it represents a recently detached slab. In the first case we would expect an active flexure-hinge retreat, accommodated by two major free boundaries at the northern and southern terminations of the arc expressed at the surface by a sinistral strike-slip fault system and by a dextral one, respectively. In effect these systems exist where they are expected (MORETTI et al., 1994; NERI et al., 1996), but we are not sure whether they presently act as major free boundaries. A serious element of doubt is represented by the absence of shallow earthquakes with thrust mechanisms in the Calabrian-Arc Ionian offshore, where they would be expected in the case of a persisting flexure-retreat. On the other hand, the major extensional features on land, from the Crati to the Mesima valleys, closely follow the elongation of the mountain belt and the reconstructed stress field shows extension axes normal to the dip of the sinking slab (GUERRA *et al.*, 1981; MONACO *et al.*, 1996; TORTORICI *et al.*, 1995). *T* axes with NE-SW direction, on the contrary, would be expected in the event that Calabria underwent the same evolution as the Southern Apennines. In conclusion, we prefer the first hypothesis, although we cannot exclude that the Calabrian Arc is presently experiencing the same change of geodynamic regime which took place in the Southern Apennines 650 Kiloyears ago. Following this choice, we consider the longitudinal extensional structures of Calabria as the equivalent of the extensional features in the axial belt of the Northern Apenninic Arc. In the hypothesis that the rate of flexure-hinge retreat has not remarkably decreased with respect to Lower-Middle Pleistocene times, we should expect in Calabria slip rates considerably higher than in the Northern Apennines, in accordance with the historical seismicity of the region.

The seismotectonic interpretation of the Calabrian Arc obviously influences the interpretation of Northern Sicily where a dextral transpressional shear zone is expected, compatible with the seismicity pattern of the area.

In Southern Sicily the geological picture is quite different, the active-fault pattern being dominated by the first-order NNW-SSE Malta fault-system and by a second-order NE-SW fault system related to the flexure of the Hyblean Plateau. An important free-boundary in Western Sicily accommodates the differential foreland flexure-retreat, maximum at the northwestern margin of the Hyblean Plateau and minimum in the Sciacca zone. We attribute the seismicity of Western Sicily (see 1968 Belice earthquake, $M = 5.9$) to the activity of this tear fault.

The active tectonics of the Gargano-Tremiti and Central Adriatic region is still poorly understood, in spite of the rich information coming from off-shore oil exploration. An inversion active tectonics (ARGNANI *et al.*, 1993; FAVALI *et al.*, 1993), with remobilization of previous extensional faults in a compressional/transpressional regime, seems to be a likely working hypothesis.

Returning to the seismotectonic model of the entire national territory and adjacent areas, we see that the seismicity pattern is controlled within a relatively small space by a very complex geodynamic framework. Within this framework we can recognize:

– continent–continent convergence (Alps and Dinarides) with development of compressive-transpressive features along the plate margins;
– plate divergence across margins characterized by passive slab sinking (Northern Apennines and Calabrian Arc), with development of backarc basins (Northern Tyrrhenian Sea and Southern Tyrrhenian Sea) flanked by forelandward migrating thrust belt-foredeep systems;

– plate divergence across a margin previously characterized by lithosphere sinking and thereafter discharged from the subducted slab (Southern Apennines), with development of quite peculiar rift processes within the inactive thrust belt;

– transpression (Northern Sicily) due to the combined effect of plate convergence (Africa-Europe) and high-rate flexure-hinge retreat of an intervening plate (Adria microplate) with high angles between the respective slip vectors;

– intraplate strain partition and fault activity (mainly combined strike-slip and thrust motions), possibly in correspondence to inverted structures.

Within the single mobile belts the earthquake space distribution and the maximum source dimensions are obviously controlled by the overall geometry of

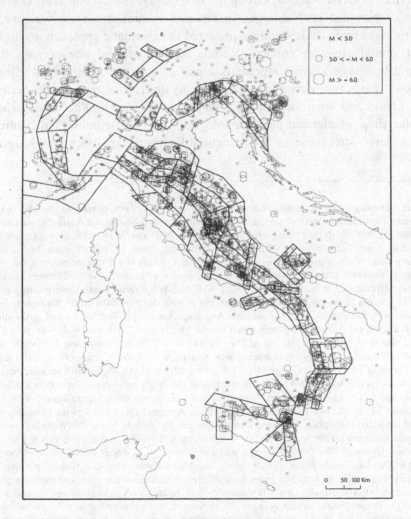

Figure 8
Seismic zonation of Italy and earthquake epicenters according to the NT Catalogue (CAMASSI and STUCCHI, 1996).

the system (3-D structural model), while the slip rate and the focal mechanisms are determined by the kinematics of the mobile lithosphere-asthenosphere system (kinematic model). Figures 8 and 9 and relative captions summarize the present-day state of the art regarding the seismic zonation of Italy.

5. Concluding Remarks

We wish to underline that the zonation described is subject to periodic revisions and improvements. A new version is currently in progress within the framework of the activities of CNR-National Group for Defense against Earthquakes. Due to the absence of standard methodologies in regional seismotectonic analyses, we were forced to make basic choices and we preferred the kinematic approach, owing to its implicit multidisciplinary content. Results have reinforced our opinion that a structural/kinematic analysis represents a useful tool to reach the first seismotectonic goal, that is to understand where and why destructive earthquakes occur in a certain region and what kind of mechanisms are expected. Other approaches, in particular those of classical geomorphological and paleoseismological nature, are not considered alternative to the structural/kinematic approach but integrative,

---

Figure 9

Kinematic behaviour of the seismic source zones of Italy. a. Zones related to the Adria-Europe convergence. Expected mechanisms: prevailing thrust with *P* axes following the Adria slip vectors (zones 4, 6, 8, 16–21); NW-SE dextral transpression (zones 1–3); W-E and WNW-ESE dextral (zones 10, 15) and sinistral (zone 22) strike-slip; N-S sinistral strike-slip (zone 5); mixed thrust and strike-slip mechanisms (zone 9). b. Alps-Apennine transfer zones and Ligurian Sea. Expected mechanisms: sinistral strike-slip in shallower crustal structures and dip-slip in deeper crustal structures (zones 23, 25, 26); compression (thrust and sinistral strike-slip with W-E and WNW-ESE *P* axes) overprinting previous extensional features (zone 24). c. Zones related to the passive sinking of the Adria lithosphere beneath the mountain chain in the Northern Apenninic Arc. Expected mechanisms: thrust and strike-slip with SW-NE *P* axes in the Adriatic longitudinal belt (zones 30, 35, 38, 48, 53); mostly dip-slip with SW-NE *T* axes in the axial belt (zones 28, 29, 32–34, 36, 37, 44–47, 50–52); prevailing NNE-SSW dextral strike-slip and subordinate dip-slip (deeper crustal structures) in transfer zones (40, 55); dip-slip with SW-NE *T* axes in the Tyrrhenian longitudinal belt (zones 27, 31, 41, 42, 49, 54) with possible NNE-SSW dextral strike-slip. d. Zones related to the deactivation of the thrust belt-foredeep system in the Southern Apennines and to the counterclockwise rotation of Adria. Expected mechanisms: dip-slip with SW-NE *T* axes (zones 57, 58, 62–64). e. Zones of the Calabrian Arc, probably related to the persisting passive sinking of the Adria lithosphere. Expected mechanisms: dip-slip with W-E and WNW-ESE *T* axes in the longitudinal structures (zones 66, 67, 69–72); W-E sinistral strike-slip (zones 65, 68); WNW-ESE dextral transpression (zones 75, 76); NW-SE dextral strike-slip (zone 74). f. Zones related to the Africa-Adria divergence. Expected mechanisms: dip-slip along the Malta Escarpment and strike-slip along minor transfer faults (zone 79). g. Foreland zones with different kinematic behaviours. Expected mechanisms: flexure-related NE-SW dip-slip (zone 78); transfer-fault related N-S dextral strike-slip and possibly dip-slip in the deeper crustal structures (zone 77); thrust and strike-slip with *P* axes following the Adria slip vectors (zones 7 and 59–61). h. Zones in active volcanic areas. Expected mechanisms: dip-slip in the Ischia-Phlegrean Fields and Vesuvius region (zone 56); dip-slip and NW-SE dextral strike-slip in the Etna region (zone 73).

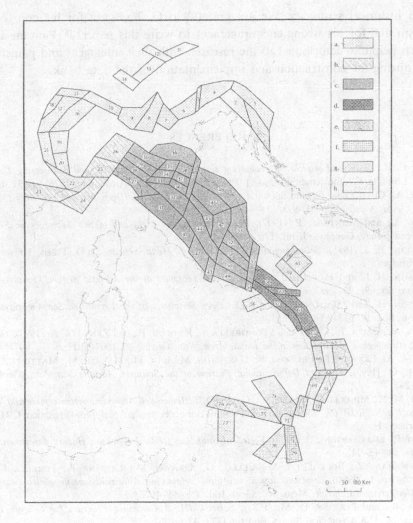

Fig. 9.

since they contribute to the characterization of the single seismic sources in terms of maximum expected magnitude, earthquake recurrence, etc. Obviously this is only a point of view.

Acknowledgements

We wish to thank all the researchers of the Working Group on Seismotectonics of the CNR-National Group for Defense against Earthquakes who contributed to the seismotectonic model and to the seismic zonation of Italy. We thank M. Stucchi for his constructive criticism and for sharing with us his deep knowledge of the

Italian historical seismicity. We are grateful to G. F. Panza for his constructive criticism and for his strong encouragement to write this paper. P. Pantani and S. Ruberti helpfully contributed to the research with their intelligent and painstaking work during the construction and implementation of the data bank.

REFERENCES

ALLEN, C. R., *Geological criteria for evaluating seismicity.* In *Seismic Risk and Engineering Decisions* (eds. Lomnitz, C., and Rosenblueth, C.) (Elsevier Sci. Publ. Co., Amsterdam 1976) pp. 31–69.

AMATO, A., CHIARABBA, C., and SELVAGGI, G. (1997), *Crustal and Deep Seismicity in Italy (30 years after)*, Ann. Geofis. *40*, 981–993.

AMATO, A., and MONTONE, P. (1997), *Present-day Stress Field and Active Tectonics in Southern Peninsular Italy*, Geophys. J. Int. *130*, 519–534.

ANDERSON, H. J. (1985), *Seismotectonics of the Western Mediterranean*, Ph.D. Thesis, University of Cambridge.

ANDERSON, H. J., and JACKSON, J. (1987a), *Active Tectonics of the Adriatic Region*, Geophys. J. R. Astron. Soc. *91*, 937–983.

ANDERSON, H., and JACKSON, J. (1987b), *The Deep Seismicity of the Tyrrhenian Sea*, Geophys. J. R. Astron. Soc. *91*, 613–637.

ANZIDEI, M., BALDI, P., CASULA, G., PONDRELLI, S., RIGUZZI, F., and ZANUTTA, A. (1997), *Geodetic and Seismological Investigation in the Ionian Area*, Ann. Geofis. *40*, 1007–1017.

ARGNANI, A., FLAVALI, P., FRUGONI, F., GASPERINI, M., LIGI, M., MARANI, M., MATTIETI, G., and MELE, G. (1993), *Foreland Deformational Pattern in the Southern Adriatic Sea*, Ann. Geofis. *36*, 229–247.

BARCHI, M. R., MINELLI, G., and PIALLI, G. (1996), *Tettonica dell'Appennino settentrionale alla luce dei risultati del CROP 03*, Soc. geol. ital. Presentazione dei risultati del profilo sismico CROP 03, Abstracts, 11–12.

BASHAM, P., and GIARDINI, D. (1993), *Technical Guidelines for Global Seismic Hazard Assessment*, Ann. Geofis. *36*, 15–24.

BEN AVRAHAM, Z., BOCCALETTI, M., CELLO, G., GRASSO, M., LENTINI, F., TORELLI, L., and TORTORICI, L. (1990), *Principali domini strutturali originatiesi dalla collisione neogenico-quaternaria nel Mediterraneo Centrale*, Mem. Soc. Geol. Ital. *45*, 453–462.

BENDER, B., and PERKINS, D. M. (1987), *Seisrisk III: A Computer Program for Seismic Hazard Estimation*, U.S. Geological Survey Bulletin 1772, 48 pp.

BOCCALETTI, M., and DAINELLI, P. (1982), *Il sistema regmatico neogenico-quatermario nell'area mediterranea: esempio di deformazione plastico-rigida post-collisionale*, Mem. Soc. Geol. Ital. *24*, 465–482.

BONCIO, P., PONZIANI, F., BROZZETTI, F., BARCHI, M., LAVECCHIA, G., and PIALLI, G. (1996), *Analisi sismotettonica dell'area compresa fra la Valle del Tevere ed il fronte della catena appenninica*, Soc. geol. ital. Presentazione dei risultati del profilo sismico CROP 03, Abstracts, 18–19.

BORISOV, B. A., REISNER, G. I., and SHOLPO, V. N. (1976), *Tectonics and Maximum Magnitudes of Earthquakes*, Tectonophysics *33*, 167–185.

BOSCHI, E., FERRARI, G., GASPERINI, P., GUIDOBONI, E., SMRIGLIO, G., and VALENSISE, G. (eds.) (1995a), *Catalogo dei forti terremoti in Italia dal 461 a.C. al 1980*, ING-SGA, Bologna, 970 pp.

BOSCHI, E., GASPERINI, P., and MULARGIA, F. (1995b), *Forecasting where Larger Crustal Earthquakes Are Likely to Occur in Italy in the Near Future*, Bull. Seismol. Soc. Amer. *85* (5), 1475–1482.

BOSCHI, E., GUIDOBONI, E., FERRARI, G., VALENSISE, G., and GASPERINI, P. (eds.) (1997), Catalogo dei forti terremoti in Italia dal 461 a.C. al 1990, ING-SGA, Bologna, 644 pp.

BRUNAMONTE, F., MICHETTI, A. M., SERVA, L., and VITTORI, E., *Evidenze paleosismologiche nell'Appennino centrale ed implicazioni neotettoniche.* In *Studi preliminari all'acquisizione dati del profilo*

CROP 11 Civitavecchia-Vasto (eds. Tozzi, M., Cavinato, G. P., and Parotto, M.) (AGIP-CNR-ENEL, Stud. geol. Camerti, vol. spec., Camerino 1991) pp. 265–270.

BUNE, V. I., MEDVEDEV, S. V., RIZNICHENKO, Yu. V., and SHEBALIN, H. V. (1974), Successes in and Expectations from the Seismic Zoning of the USSR, Izv. Earth Phys. (10), 95–102.

CAMASSI, R., and STUCCHI, M. (1996), NT4.1, un catalogo parametrico di terremoti di area italiana al di sopra della soglia del danno, CNR-GNDT. Internet: http://emidius.itim.mi.cnr.it/NT/home.html

CASTELLARIN, A. (1978), Il problema dei raccorciamenti crostali nel Sudalpino, Rend. Soc. Geol. Ital. 1, 21–23.

CELLO, G., MAZZOLI, S., TONDI, E., and TURCO, E. (1997), Active Tectonics in the Central Apennines and Possible Implications for Seismic Hazard Analysis in Peninsular Italy, Tectonophysics 272, 43–68.

CHASE, C. G. (1978), Plate Kinematics: The Americas, East Africa and the Rest of the World, Earth Planet. Sci. Lett. 37, 355–368.

CHIARABBA, C., and AMATO, A. (1996), Crustal Velocity Structure of the Apennines (Italy) from P-wave Travel Time Tomography, Ann. Geofis. 39, 1133–1148.

CINQUE, A., PATACCA, E., SCANDONE, P., and TOZZI, M. (1993), Quaternary kinematic evolution of the Southern Apennines. Relationships between surface geological features and deep lithospheric structures, Spec. Issue on the Workshop: "Modes of crustal deformation: from the brittle upper crust through detachments to the ductile lower crust" (Erice, 18–24 November 1991), Ann. Geofis. 36, 249–260.

CINTI, F. R., CUCCI, L., PANTOSTI, D., D'ADDEZIO, G., and MEGHRAOUI, M. (1997), A Major Seismogenetic Fault in a "Silent Area": The Castrovillari Fault (Southern Apennines, Italy), Geophys. J. Int. 130, 322–332.

CISTERNAS, A., GODEFROY, P., GVISHIANI, A., GORSHKOV, A. I., KOSOBOKOV, V., LAMBERT, M., RANZMAN, E., SALLANTIN, J., SOLDANO, H., SOLOVIEV, A., and WEBER, C. (1985), A Dual Approach to Recognition of Earthquake Prone Areas in the Western Alps, Ann. Geophys. 3, 249–270.

CNR, P.F. GEODINAMICA (1990), Structural Model of Italy 1:500.000 and Gravity Map, Quad. Ric. Sci., 3 (114), S.EL.CA., Firenze.

CORNELL, C. A. (1968), Engineering Seismic Risk Analysis, Bull. Seismol. Soc. Am. 58, 1583–1606.

COSTA, G., OROZOVA-STANISHKOVA, I., PANZA, G. F., and ROWAIN, I. M. (1996), Seismotectonic Models and CN Algorithm: The Case of Italy, Pure appl. geophys. 147, 119–130.

COSTA, G., PANZA, G. F., and ROTWAIN, I. M. (1995), Stability of Premonitory Seismicity Pattern and Intermediate-term Earthquake Prediction in Central Italy, Pure appl. geophys. 145 (2), 259–275.

COSTA, G., PANZA, G. F., SUHADOLC, P., and VACCARI, F. (1993), Zoning of the Italian Territory in Terms of Expected Peak Ground Acceleration Derived from Complete Synthetic Seismograms, J. Appl. Geophys. 30, 149–160.

DE METS, C., GORDON, R. G., ARGUS, D. F., and STEIN, S. (1990), Current Plane Motions, Geophys. J. Int. 101, 425–478.

DEWEY, J. F., HELMAN, M. L., TURCO, E., HUTTON, D. H. W., and KNOTT, S. D. (1989), Kinematics of the western Mediterranean. In Alpine Tectonics (eds. Coward, M. P., Dietrich, D., and Park, R. G) (Geol. Soc. Spec. Publ. 45) pp. 265–283.

DEWEY, J. F., PITMAN, W. C., RYAN, W. B. F., and BONNIN, J. (1973), Plate Tectonics and the Evolution of the Alpine System, Bull. Geol. Soc. Am. 84, 3137–3180.

D'OFFIZI, S. (1994), Overall Seismotectonic Modelling—A Powerful Tool in Seismic Hazard Analyses: Two Applications in Central Italy and California (U.S.A.), Boll. Geofis. Teor. Appl. 36 (141–144), 331–361.

DOGLIONI, C. (1991), A Proposal of Kinematic Modelling for W-dipping Subductions—Possible Applications to the Tyrrhenian-Apennine System, Terra Nova 3, 423–434.

DOGLIONI, C., HARABAGLIA, P., MARTINELLI, G., MONGELLI, F., and ZITO, G. (1996), A Geodynamic Model of the Southern Apennines Accretionary Prism, Terra Nova 8, 540–547.

DZIEWONSKI, A. M., FRIEDMAN, A., GIARDINI, D., and WOODHOUSE, J. H. (1983), Global Seismicity of 1982: Centroid Moment Tensor Solutions for 308 Earthquakes, Phys. Earth Planet. Inter. 53, 17–45.

EVA, E., SOLARINO, S., and EVA, C. (1997), Stress Tensor Orientation Derived from Fault Plane Solutions in the Southwestern Alps, J. Geophys. Res. 102, 8171–8185.

FAVALI, P., FUNICIELLO, R., MATTIETTI, G., MELE, G., and SALVINI, F. (1993), An Active Margin across the Adriatic Sea (Central Mediterranean Sea), Tectonophysics 219, 109–117.

FERRELI, L., MICHETTI, A. M., SERVA, L., VITTORI, E., and ZAMBONELLI, E. (1996), *Tettonica recente ed evidenze di fagliazione superficiale nella catena del Pollino (Calabria settentrionale)*, Mem. Soc. Geol. Ital. *51*, 451–466.

FINETTI, I., and DEL BEN, A. (1986), *Geophysical Study of the Tyrrhenian Opening*, Boll. Geofis. Teor. Appl. *28*, 75–155.

GALADINI, F., GALLI, P., and GIRAUDI, C. (1996), *Paleosismologia della Piana del Fucino (Italia Centrale)*, II Quaternario *10* (1), 27–64.

GASPARINI, C., IANNACCONE, G., SCANDONE, P., and SCARPA, R (1982), *Seismotectonics of the Calabrian Arc*, Tectonophysics *84*, 267–286.

GELFAND, I. M., GUBERMAN, SH. I., IZVEKOVA, M. L., KELIS-BOROK, V. I., and RANZMAN, E. JA. (1972), *Criteria of High Seismicity, Determined by Pattern Recognition*, Tectonophysics *13*, 415–422.

GHISETTI, F. (1992), *Fault Parameters in the Messina Strait (Southern Italy) and Relations with the Seismogenic Source*, Tectonophysics *210*, 117–133.

GIARDINI, D., and BASHAM, P. (1993), *The Global Seismic Hazard Assessment Program (GSHAP)*, Ann. Geofis. *36*, 3–13.

GIARDINI, D., and VELONÀ, M. (1991), *The Deep Seismicity of the Tyrrhenian Sea*, Terra Nova *3*, 57–64.

GRELLET, B., COMBES, P., GRANIER, T., PHILIP, H., MOHAMMADIOUN, B. (1993), *Sismotectonique de la France métropolitaine, dans son cadre géologique et géophysique*, Mem. Soc. Geol. France, *164*, 2 vol., 76 pp.

GRÜNTHAL, G., BOSSE, C., MAYER-ROSA, D., RÜTTENER, E., LENHARDT, W., and MELICHAR, P. (1995), *Across-boundaries Seismic Hazard Maps in the GSHAP-Region 3-Case Study for Austria, Germany and Switzerland*, Proc. of the 24th Assembly ESC (Athens) 19–24 Sept. 1994, 1542–1548.

GUERRA, I., SCARPA, R., TORTORICI, L., and TURCO, E. (1981), *Geometria del campo tensionale agente in Calabria settentrionale: confronti tra metodologie strutturali e sismologiche*, Rend. Soc. Geol. Ital. *4*, 109–112.

HAYS, W. W. (1980), *Procedures for Estimating Earthquake Ground Motion*, Professional Paper 1114, United States Geophysical Survey, Washington, 77 pp.

HELMAN, M. L. (1989), *Tectonics of the Western Mediterranean*, Ph.D. Thesis, University of Oxford.

HERAK, M., HERAK, D., and MARKUSIC, S. (1995), *Fault-plane Solutions (1956–1995) in Croatia and Neighbouring Regions*, Geofizika *12*.

HIPPOLYTE, J. C., ANGELIER, J., and ROURE, F. (1994), *A Major Geodynamic Change Revealed by Quaternary Stress Patterns in the Southern Apennines (Italy)*, Tectonophysics *230*, 199–210.

HIRN, A., NICOLICH, R., GALLART, J., LAIGLE, M., CERNOBORI, L., and GROUP, E. S. (1997), *Root of Etna Volcano in Faults of Great Earthquakes*, Earth Planet. Sci. Lett. *148*, 171–191.

IAEA SAFETY GUIDE (1991), *Earthquake and Associated Topics in Relation to Nuclear Power Plant Siting*, International Atomic Energy Agency, Safety Series, 50-SG-S1 (Rev. 1), Vienna, 49 pp.

ING (1995), *Seismological Reports 1980–1995*, Istituto Nazionale di Geofisica, Roma, Italy (computer file).

LAVECCHIA, G., BROZZETTI, F., BARCHI, M., MENICHETTI, M., and KELLER, J. V. A. (1994), *Seismotectonic Zoning in East-central Italy Deduced from an Analysis of the Neogene to Present Deformations and Related Stress Fields*, Bull. Geol. Soc. Am. *106*, 1107–1120.

LENHARDT, W. A. (1995), *Earthquake Hazard in Austria*, Proc. of 24th Assembly ESC (Athens) 19–24 Sept. 1994, 1499–1507.

LENTINI, F., CARBONE, S., and CATALANO, S. (1996), *The External Thrust System in Southern Italy: A Target for Petroleum Exploration*, Petroleum Geosci. *2*, 33–342.

LIVERMORE, R. A., and SMITH, A. G., *Some boundary conditions for the evolution of the Mediterranean region*. In *Geological Evolution of the Mediterranean Basin*, Raimondo Selli Commemorative Vol. (eds. Stanley, D. J., and Wezel, F. C.) (Springer-Verlag, New York 1985) pp. 83–98.

MALINVERNO, A., and RYAN, W. B. F. (1986), *Extension in the Tyrrhenian Sea and Shortening in the Apennines as a Result of Arc Migration Driven by Sinking of the Lithosphere*, Tectonics *5*, 227–245.

MANTOVANI, E., ALBARELLO, D., TAMBURELLI, C., and BABBUCCI, D. (1996), *Evolution of the Tyrrhenian Basin and Surrounding Regions as a Result of the Africa-Eurasia Convergence*, J. Geodyn. *21* (1), 35–72.

MARSON, I., PANZA, G. F., and SUHADOLC, P. (1995), *Crust and Upper Mantle Models along the Active Tyrrhenian Rim*, Terra Nova 7, 348–357.

MAZZOLI, S., and HELMAN, M. (1994), *Neogene Patterns of Relative Plate Motions for Africa-Europe: Some Implications for Recent Central Mediterranean Tectonics*, Geol. Rundsch. 83, 464–468.

McKENZIE, D. (1972), *Active Tectonics of the Mediterranean Region*, Geophys. J. R. Astron. Soc. 30, 109–185.

MELETTI, C., PATACCA, E., and SCANDONE, P., *Il sistema compressione-distensione in Appennino*. In *Cinquanta anni di attiviatà didattica e scientifica del Prof. Felice Ippolito* (eds. Bonardi, G., De Vivo, B., Gasparini, P., and Vallario, A.) (Liguori, Napoli 1995) pp. 361–370.

MICHETTI, A. M., BRUNAMONTE, F., SERVA, L., and VITTORI, E. (1996), *Trench Investigations of the 1915 Fucino Earthquake Fault Scarps (Abruzzo, Central Italy): Geological Evidence of Large Historical Events*, J. Geophys. Res. 101, 5921–5936.

MINSTER, J. B., and JORDAN, T. H. (1978), *Present-day Plate Motions*, J. Geophys. Res. 83, 5331–5354.

MOLIN, D., STUCCHI, M., and VALENSISE, G. (1996), *Massime intensità macrosismiche osservate nei comuni italiani*, GNDT-ING-SSN Report for Civil Defense Department. Available on Internet: http://emidius.itim.mi.cnr.it/GNDT/IMAX/max_int_oss.html

MONACHESI, G., and STUCCHI, M. (1997), *DOM4.1, un database di osservazioni macrosismiche di terremoti di area italiana al di sopra della soglia del danno*, CNR-GNDT. Internet: http://emidius.itim.mi.cnr.it/DOM/home.html

MONACO, C., TORTORICI, L., NICOLICH, R., CERNOBORI, L., and COSTA, M. (1996), *From Collisional to Rifted Basins: An Example from the Southern Calabrian Arc (Italy)*, Tectonophysics 266, 233–249.

MONTONE, P., AMATO, A., FREPOLI, A., MARIUCCI, M. T., and CESARO, M. (1997), *Crustal Stress Regime in Italy*, Ann. Geofis. 40, 741–757.

MORETTI, A., CURRÀ, M. F., and GUERRA, I. (1994), *Shallow Seismicity Activity in the Southeastern Tyrrhenian Sea and along the Calabria-Lucania Border, Southern Italy*, Boll. Geofis. Teor. Appl. 36 (141–144), 399–409.

MUCO, B. (1994), *Focal Mechanism Solutions for Albanian Earthquakes for the Years 1964–1988*, Tectonophysics 231, 311–323.

MUIR-WOOD, R. (1993), *From Global Seismotectonics to Global Seismic Hazard*, Ann. Geofis. 36, 153–168.

MULARGIA, F., GASPERINI, P., and TINTI, S. (1987), *A Procedure to Identify Objectively Active Seismotectonic Structures*, Boll. Geofis. Teor. Appl. 29 (114), 147–164.

MÜLLER, B., ZOBACK, M. L., FUCHS, K., MASTIN, L., GREGERSEN, S., PAVONI, N., STEPHANSSON, O., and LJUNGGREN, C. (1992), *Regional Patterns of Tectonic Stress in Europe*, J. Geophys. Res. 97, 11783–11803.

NERI, G., CACCAMO, D., COCINA, O., and MONTALTO, A. (1996), *Geodynamic Implications of Earthquake Data in the Southern Tyrrhenian Sea*, Tectonophysics 258, 233–249.

NICOLICH, R. (1989), *Crustal Structures from seismic studies in the frame of the European geotraverse (southern segment) and CROP projects*. In *The Lithosphere in Italy. Advances in Earth Science Research. It. Nat. Comm. Int. Lith. Progr., Mid Term Conf. (Rome, 5–6 May 1987)* (eds. Boriani, A., Bonafede, M., Piccardo, G. B., and Vai, G. B.), Atti Conv. Lincei 80, 41–61.

PANTOSTI, D., D'ADDEZIO, G., and CINTI, F. R. (1996), *Paleoseismicity of the Ovindoli-Pezza Fault, Central Apennines, Italy: A History Including a Large, Previously Unrecorded Earthquake in Middle Age (860–1300 A.D.)*, J. Geophys. Res. 101, 5937–5959.

PANTOSTI, D., SCHWARTZ, D. P., and VALENSISE, G. (1993), *Paleoseismology along the 1980 Surface Rupture of the Irpinia Fault: Implications for the Earthquake Recurrence in the Southern Apennines, Italy*, J. Geophys. Res. 98, 6561–6577.

PANTOSTI, D., and YEATS, R. S. (1993), *Paleoseismology of Great Earthquakes of the Late Holocene*, Ann. Geofis. 36, 237–257.

PANZA, G. F., VACCARI, F., and CAZZARO, R., *Deterministic seismic hazard assessment*. In *Vrancea Earthquakes: Tectonics, Hazard and Risk Mitigation* (eds. Wenzel, F., and Lungu, D.) (Kluwer Academic Publ, Dordrecht 1998) pp. 269–286.

PAPADIMITRIOU, E. E. (1993), *Focal Mechanism along the Convex Side of the Hellenic Arc*, Boll. Geofis. Teor. Appl. 35 (140), 402–426.

PAPAZACHOS, B. C. (1988), *Seismic Zones in Aegean and Surrounding Areas*, Proc. 21st General Assembly ESC (Sofia 1988) 1–6.

PAPAZACHOS, B. C. (1996), *Large Seismic Faults in the Hellenic Arc*, Ann. Geofis. *39*, 891–903.

PATACCA, E., SARTORI, R., and SCANDONE, P. (1990), *Tyrrhenian Basin and Apenninic Arcs: Kinematic Relations since Late Tortonian Times*, Mem. Soc. Geol. Ital. *45*, 425–451.

PATACCA, E., SARTORI, R., and SCANDONE, P., *Tyrrhenian basin and Apennines. Kinematic evolution and related dynamic constraints*. In *Recent Evolution and Seismicity of the Mediterranean Region* (eds. Boschi, E., Mantovani, E., and Morelli, A.) (Kluwer Academic Publ., Dordrecht 1993) pp. 161–171.

PATACCA, E., and SCANDONE, P. (1986), *Seismical Hazard: Seismotectonic Approach*, IAEG Int. Symp. on *Engineering Geology Problems in Seismic Areas* (Bari, 13–19 April 1986), *5*, 103–115.

PATACCA, E., and SCANDONE, P. (1989), *Post-Tortonian mountain building in the Apennines. The role of the passive sinking of a relic lithospheric slab*. In *The Lithosphere in Italy. Advances in Earth Science Research. It. Nat. Comm. Int. Lith. Progr., Mid-term Conf. (Rome, 5–6 May 1987)* (eds. Boriani, A., Bonafede, M., Piccardo, G. B., and Vai, G. B.), Atti Conv. Lincei, *80*, 157–176.

PATACCA, E., SCANDONE, P., and MELETTI, C. (1997), *Variazioni di regime tettonico nell'Appennino meridionale durante il Quaternario*, Conv. AIQUA "Tettonica Quaternaria del Territorio Italiano: Conoscenze, Problemi ed Applicazioni" (Parma, 25–27 Feb. 1997), Riassunti.

PERUZZA, L., PANTOSTI, D., SLEJKO, D., and VALENSISE, L. (1997), *Testing a New Hybrid Approach to Seismic Hazard Assessment: An Application to the Calabrian Arc (Southern Italy)*, Nat. Hazards *14*, 113–126.

POSTPISCHL, D. (1985), *Catalogo dei terremoti italiani dall'anno 1000 al 1980*. Quaderni della Ricerca Scientifica, 114, 2B, Bologna, 239 pp.

RAGG, S., GRASSO, M., and MÜLLER, B. (1995), *3-D FE Computation of Tectonic Stresses in Sicily Combined with Results of Breakout Analysis*, Terra Nova 7 (abstract suppl. 1), 170.

RENNER, G., and SLEJKO, D. (1994), *Some Comments on the Seismicity of the Adriatic Region*, Boll. Geofis. Teor. Appl. *36* (141–144), 381–398.

ROMEO, R., and PUGLIESE, A. (1997), *La pericolosità sismica in Italia. Parte 1: analisi della scuotibilità*, Rapporto Tecnico SSN/RT/97/1, Roma, Gennaio 1997.

ROURE, F., HEITZMANN, P. R., and POLINO, R. (eds.) (1990), *Deep Structure of the Alps*, Mém. Soc. Géol. France, 156; Mém. Soc. Géol. Suisse, *1*, Vol. Spec. Soc. Geol. Ital., *1*, 367 pp.

ROYDEN, L., PATACCA, E., and SCANDONE, P. (1987), *Segmentation and Configuration of Subducted Lithosphere in Italy: An Important Control on Thrust-belt and Foredeep-basin Evolution*, Geology *15*, 714–717.

SARTORI, R., *The main results of ODP Leg 107 in the frame of Neogene to Recent geology of Perityrrhenian areas*. In *Proceeding Ocean Drilling Program, Sci. Results* (eds. Kastens, K. A., Mascle, J. *et al.*), *107* (Ocean Drilling Program, College Station, TX 1990) pp. 715–730.

SCANDONE, P., PATACCA, E., MELETTI, C., BELLATALLA, M., PERILLI, N., and SANTINI, U. (1992), *Struttura geologica, evoluzione cinematica e schema sismotettonico della penisola italiana*, Atti del Convegno Annuale del Gruppo Nazionale per la Difesa dai Terremoti (Pisa, 25–27 giugno 1990), 1, Ed. Ambiente, Bologna 1992, 119–135.

SCANDONE, P., PATACCA, E., RADOICIC, R., RYAN, W. B. F., CITA, M. B., RAWSON, M., CHEZAR, H., MILLER, E., MCKENZIE, J., and ROSSI, S. (1981), *Mesozoic and Cenozoic Rocks from Malta Escarpment (Central Mediterranean)*, Bull. Am. Assoc. Petroleum Geol. *65*, 1299–1319.

SCHICK, R., *Seismotectonic survey of the Central Mediterranean*. In *Alps Apennines Hellenids* (eds. Closs, H., Roeder, D., and Schmidt, K.), Inter-Union Commission on Geodynamics Scientific Report No. 38, E. Schweizerbart'sche Verlagsbuchhandlung (Nägele und Obermüller, Stuttgart 1978), 335–338.

SCHMID, S. M., PFIFFNER, O. A., FROITZHEIM, N., SCHÖNBORN, G., and KISSLING, E. (1996), *Geophysical-geological Transect and Tectonic Evolution of the Swiss-Italian Alps*, Tectonics *15*, 1036–1064.

SCHWARTZ, D. P., and COPPERSMITH, K. J., *Seismic hazards: new trends in analysis using geologic data*. In *Active Tectonics. Studies in Geophysics* (National Academy Press, Washington 1986) pp. 215–230.

SEARLE, R. (1980), *Tectonic Pattern of the Azores Spreading Centre and Triple Junction*, Earth Planet. Sci. Lett. *51*, 415–434.

SERRI, G. (1997), *Neogene-quaternary Magmatic Activity and its Geodynamic Implications in the Central Mediterranean Region*, Ann. Geofis. *40*, 681–703.

SERVA, L., BLUMETTI, A. M., and MICHETTI, A. (1986), *Gli effetti sul terreno del terremoto del Fucino (13 gennaio 1915); tentativo di interpretazione dell'evoluzione tettonica recente di alcune strutture*, Mem. Soc. Geol. Ital. *35*, 893–907.

SLEJKO, D. (1996), *Pericolosità Seismica del territorio nazionale*, CNR-GNDT, Internet: http://emidius.itim.mi.cnr.it/GNDT/home.thml

SLEJKO, D., PERUZZA, L., and REBEZ, A. (1998), *Seismic Hazard Maps of Italy*, Ann. Geofis. *41* (2), 183–214.

SLEMMONS, D. B., and DEPOLO, C. M., *Evaluation of active faulting and associated hazards.* In *Active Tectonics. Studies in Geophysics* (National Academy Press, Washington 1986) pp. 45–62.

SUHADOLC, P., MARSON, I., and PANZA, G. F. (1993), *Crust and Upper Mantle Structural Properties along the Active Tyrrhenian Rim*, Acta Geod. Geoph. Mont. Hung. *28*, 307–321.

SUHADOLC, P., and PANZA, G. F. (1989), *Physical properties of the lithosphere-asthenosphere system in Europe from geophysical data.* In *The Lithosphere in Italy. Advances in Earth Science Research. It. Nat. Comm. Int. Lith. Progr., Mid-term Conf. (Rome, 5–6 May 1987)* (eds. Boriani, A., Bonafede, M., Piccardo, G. B., and Vai, G. B.), Atti Conv. Lincei, *80*, 15–40.

TORTORICI, L., MONACO, C., TANSI, L., and COCINA, O. (1995), *Recent and Active Tectonics of the Calabrian Arc (Southern Italy)*, Tectonophysics *243*, 37–55.

TRIFONOV, V. G., and MACHETTE, M. N. (1993), *The World Map of Major Active Faults Project*, Ann. Geofis. *36*, 225–236.

VALENSISE, G., and PANTOSTI, D. (1992), *A 125 Kyr-long Geological Record of Seismic Source Repeatability: The Messina Straits (Southern Italy) and the 1908 Earthquake (M_s 7 1/2)*, Terra Nova *4*, 472–483.

VOGT, J., and GODEFROY, P. (1981), *Carte sismotectonique de la France. Presentation et mode d'emploi. Commentaire des cartouches*, Memoire BRGM, *111*, Orleans, 27 pp.

WALSH, J. J., and WATTERSON, J. (1988), *Analysis of the Relationship between Displacements and Dimensions of Faults*, J. Struct. Geol. *10*, 238–247.

WARD, S. N. (1994), *Constraints on the Seismotectonics of the Central Mediterranean from Very Long Baseline Interferometry*, Geophys. J. Int. *117*, 441–452.

WARD, S. N., and VALENSISE, G. L. (1989), *Fault Parameters and Slip Distribution of the 1915 Avezzano, Italy, Earthquake Derived from Geodetic Observations*, Bull. Seismol. Soc. Am. *79*, 690–710.

WELLS, D. L., and COPPERSMITH, K. J. (1994), *Updated Empirical Relationship among Magnitude, Rupture Length, Rupture Area, and Surface Displacement*, Bull. Seismol. Soc. Am. *84*, 974–1002.

WESTAWAY, R. (1990), *Present-day Kinematics of the Plate Boundary Zone between Africa and Europe, From the Azores to the Aegean*, Earth Planet. Sci. Lett. *96*, 392–406.

WORKING GROUP ON CALIFORNIA EARTHQUAKE PROBABILITIES (1995), *Seismic Hazards in Southern California: Probable Earthquake, 1994 to 2024*, Bull. Seismol. Soc. Am. *85*, 379–493.

ZOBACK, M. L. (1993), *Utilising in-situ Stress Data for Seismic Hazard Assessment: The World Stress Map Project's Contribution to GSHAP*, Ann. Geofis. *36*, 217–224.

(Received July 14, 1998, revised March 2, 1999, accepted March 2, 1999)

To access this journal online:
http://www.birkhauser.ch

Pure appl. geophys. 157 (2000) 37–55
0033–4553/00/020037–19 $ 1.50 + 0.20/0

Pure and Applied Geophysics

The Seismotectonic Characteristics of Slovenia

MARIJAN POLJAK,[1] MLADEN ŽIVČIĆ[2] and POLONA ZUPANČIČ[3]

Abstract—Slovenia with its neighbouring areas lies at the junction of the Alps, the Dinarides and the Pannonian basin. These belong to the three plates: Europe, Adria and Tisza. On the Slovenian territory itself converge the External Dinarides NW-SE oriented right lateral strike-slip faults, the Transdanubian Range NE-SW oriented left lateral strike-slip faults, and the Southern Alps E-W oriented thrusts. The direction of the principal stress σ_1 (azimuth = 6°, dip = 8°) is determined under the assumption of uniform stress throughout the region. Dip of the least principal stress σ_3 of 5° is consistent with the regional strike-slip regime. Listed structures form a pure shear structural mechanism on a regional scale.

In spite of geologic evidence of tectonic displacements along mentioned structures in the past, there is no surface expression of their recent activity.

The lithospheric units of the investigated area were amalgamated together during Tertiary. The seismicity is not concentrated along the primary plate boundaries but is rather spread in a broad zone along their deformed rims. The seismicity is moderate with the average depth of earthquakes in Slovenia of 6.5 km, and 9 to 20 km for stronger earthquakes ($M_{LH} > 4.2$). No surface rupture related to an earthquake has been detected to date in Slovenia.

The territory of Slovenia and its neighbouring regions has been delineated into five seismogenic areas, i.e., the areas with similar and among themselves differentiable tectonic and seismological characteristics. They are the Eastern Alps, the Southern Alps, with the Friuli region as a separate unit, the External Dinarides, and the Transdanubian Range.

Key words: Tectonics, seismicity, geodynamics, Slovenia, Dinarides, Alps, Pannonian basin.

Introduction

The aim of this paper is to integrate data from the revised earthquake catalogue, as well as published and newly-determined fault plane solutions, and to combine them with the available geological and geodetic information in order to determine some of Slovenia's seismotectonic characteristics. The geotectonic setting, fault plane solutions and distribution of earthquake hypocentres, as well as the results of geophysical and geodetic surveys, were of primary importance.

[1] Geological Survey of Slovenia, Dimičeva 14, SI-1000 Ljubljana, Slovenia.
[2] Geophysical Survey of Slovenia, Observatory, Pot na Golovec 25, SI-1000 Ljubljana, Slovenia and Slovenian Association for Geodesy and Geophysics, Kersnikova 3/II, SI-1000 Ljubljana, e-mail: mladen.zivcic@uni-lj.si
[3] Geophysical Survey of Slovenia, Kersnikova 3, SI-1000 Ljubljana, Slovenia.

Tectonic data have been compiled by using a geologic map of Slovenia (BUSER and DRAKSLER, 1990) and recent geological investigation. The seismicity of Slovenia and surrounding regions has been studied primarily on the basis of earthquake catalogues compiled from macroseismic data (SHEBALIN *et al.*, 1974; CVIJANOVIĆ, 1981; RIBARIČ, 1982). In addition to this, the catalogue prepared for seismic hazard assessment (ŽIVČIĆ *et al.*, this volume) has also been used in this study.

Recent stress regime was determined from field structural measurements (POLJAK, 1984) and by stress inversion from the fault plane solutions. The displacement rates are estimated from geodetic measurements on regional (KOLER, 1990) and local (KOLER and BREZNIKAR, 1998) scales.

On the basis of these data, the area has been divided into several zones. These represent the terrain of similar, but among themselves differentiable, tectonic and seismological characteristics.

The seismotectonics of the Slovenian territory has also been investigated and interpreted by authors from neighbouring countries (SLEJKO *et al.*, 1987; CARULLI *et al.*, 1990; DEL BEN *et al.*, 1991, etc.). All these studies included the seismic and seismotectonic zoning of Slovenian territory, and a calculation of their quantitative seismic parameters. However, the lack of exact geologic data has prevented a reliable determination of the particular seismogenic structures or seismogenic zones and their characteristics.

Tectonics

The geotectonic position of the Slovenian territory is relatively well understood within the framework of the plate tectonic model: the area south of the Periadriatic line is part of the Adriatic microplate, whose northern margin (Southern Alps—Dinarides) is highly deformed and backthrusted onto the central, less deformed part of the Adriatic microplate (Adriatic basin). The Periadriatic line itself represents its boundary to the European plate (Fig. 1).

According to the classical Kober's concept (KOBER, 1921, 1931) of the bilateral Alpine orogeny, the whole area south of the Periadriatic line belongs to the Dinarides, and the area north of it to the Alps. The Periadriatic line is, according to the author, the root zone of the two symmetrical branches of the Alpine orogeny. Contrary to this, numerous Alpine geologists (LEMOIN, ed., 1978) label the area south of the Periadriatic line as the Southern Alps with no clear boundary to the Dinarides. Recently, BUSER (1989) proposed clarification of these terms, at least for the Slovenian territory, and named the area south of the Periadriatic line the Dinarides "sensu lato" with a subdivision into the Southern Alps, the Internal Dinarides and the External Dinarides. On the basis of the structural pattern, we have divided the area south of the Periadriatic line into the Southern Alps and the External Dinarides.

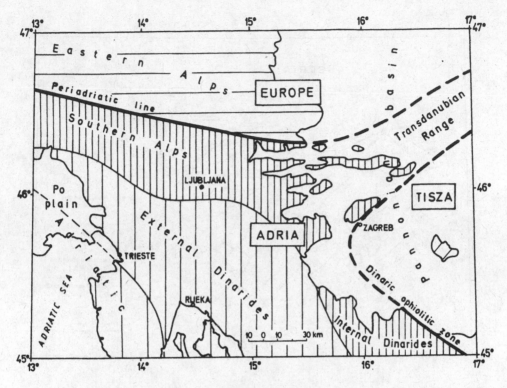

Figure 1
Geotectonic position of the area under investigations. Rocks of particular geotectonic units are in the Pannonian basin and the Po plain covered by the molasse sediments of mostly Neogene age. Names of the tectonic plates are in boxes.

The eastern limit of the Eastern and the Southern Alps towards the Pannonian basin has also not been definitely determined. Rocks of these two units form the basement of mostly Neogene sediments of the western rim of the Pannonian basin. In order to avoid further confusion, we use the term Pannonian basin for the area covered by molasse type sediments of Neogene age.

The main structural pattern of Slovenian territory is seen on Figure 2, which represents a simplified tectonic map derived from the Geologic Map of Slovenia in scale 1:500,000 by BUSER and DRAKSLER (1990). Faults labeled 1 to 7, as well as thrusts 17 to 19 belong to the External Dinarides. Thrusts 16, 20 and 21 belong to the Southern Alps, faults 9 to 11 define the Periadriatic line, and faults 12 to 15 represent the southeastern continuation of the Transdanubian Range of Central Hungary.

The External Dinarides NW-SE oriented faults have generally right-lateral horizontal displacement (PLENIČAR, 1969), as do the faults accompanying the

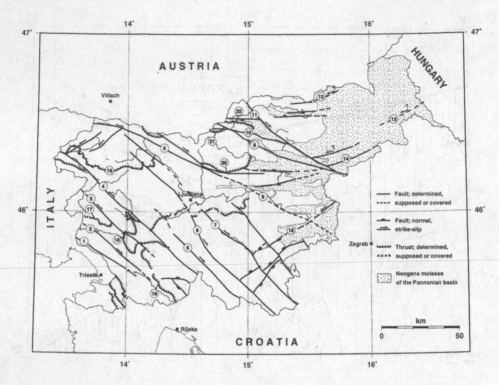

Figure 2

Tectonic map of Slovenia, generalised after BUSER and DRAKSLER (1990); Main faults: 1. Divača, 2. Raša, 3. Predjama, 4. Idrija, 5. Borovnica, 6. Želimlje, 7. Žužemberk, 8. Sava, 9. Šoštanj, 10. Smrekovec, 11. Labod, 12. Kungota, 13. Ljutomer, 14. Donat, 15. Orlica; Main thrusts: 16. Julian Alps, 17. Trnovski Gozd, 18. Nanos, 19. Snežnik, 20. Kamnik-Savinja Alps, 21. Southern Karavanke, 22. Northern Karavanke. Faults labeled 1 to 7 as well as thrusts 17 to 19 belong to the External Dinarides. Thrusts 16, 20 and 21 belong to the Southern Alps, faults 9 to 11 define the Periadriatic line, and faults 12 to 15 represent the southeastern continuation of the Transdanubian Range of Central Hungary.

Periadriatic line (BEMMELEN, 1970). The faults of the Pannonian basin have left-lateral horizontal displacement (RUMPLER and HORVÁTH, 1988).

However, geological evidence of horizontal movements along the mentioned faults, especially in Quaternary sediments is not distinct. Detailed studies of Quaternary river deposits along the Sava fault (ŽLEBNIK, 1971; KUŠČER, 1990) did not detect any structural deformations in sediments of Middle to Upper Pleistocene and Holocene age. Indirect proofs of recent fault activity in the form of rhomboidal karst fields developed along the Idrija fault have been proposed by POLJAK (1986), and by PLACER (1996b) for the intramountain depressions along the Sava fault filled up by Quaternary river deposits.

Seismicity

The catalogue prepared for seismic hazard assessment (ŽIVČIĆ *et al.*, this volume) was used in this study. It was prepared by critically merging the Slovene catalogue with the relevant parts of the national catalogues of neighbouring countries. Magnitudes of historical earthquakes were determined from macroseismic data utilizing the regression derived from the earthquakes with known magnitude M_{LH} as defined by KARNÍK (1968) and also used in SHEBALIN *et al.* (1974). Recently published data on particular earthquake sequences were also included. Earthquake locations were evaluated mostly from macroseismic data, as the number and distribution of seismological stations was insufficient for a reliable estimation of earthquake parameters until the late 1980s. The accuracy of epicentre determination is estimated to about 0.1 to 0.2 degrees (RIBARIČ, 1982). Focal depths are determined from isoseismal radii using Köveslighety's attenuation law (ŽIVČIĆ, 1984).

The territory of Slovenia can be considered to be one of moderate seismicity. The activity rate $A_{3.8}$ is defined as a number of earthquakes of magnitude 3.8 and over in the period after 1870, for which the catalogue is considered to be reasonably complete (see Fig. 3), per year per 1000 km^2. The average value for entire Slovenia is 0.09 earthquakes per year per 1000 km^2. The oldest event in the catalogue of earthquakes in Slovenia (RIBARIČ, 1982) dates back to 792, and is of still unconfirmed reliability (CECIĆ and ŽIVČIĆ, 1996). Subsequently, there has been one event of maximum intensity X MSK, two of intensity IX MSK and 11 that reached the maximum intensity VIII MSK. Only three earthquakes of intensity VIII MSK have occurred in the last 200 years.

The strongest event known to have happened on the territory of Slovenia is the so-called Idrija earthquake of 26 March 1511 (RIBARIČ, 1979). Its exact location and nature, and its possible relation to an almost simultaneous event in Friuli (NE Italy), are still uncertain. It caused extensive damage throughout Slovenia and neighbouring countries. The magnitude M_{LH} of the event was estimated, from its macroseismic effects, to have been between 6.8 (ŽIVČIĆ and CECIĆ, 1998) and 7.2 (RIBARIČ, 1979). This is the only event of intensity X MSK ever to have taken place in Slovenia.

The seismicity in Slovenia has a rather smeared pattern. This can partly be attributed to the fact that the majority of epicentres was determined from macroseismic data—they reflect the surface manifestations of an earthquake rather than its hypocentral position at depth, and are not the best for studying seismotectonic processes, except for very general features. For more detailed mapping of active structures one should use hypocentral determinations from the dense network of seismic stations which for Slovenia exists only for the last decade. Epicentres of earthquakes with a magnitude of $M_{LH} = 4.5$ and larger after 1870 are plotted on Figure 4. The distribution indicates the existence of several areas of increased seismicity.

The average depth of all earthquakes in Slovenia is 6.5 km. The depths of stronger earthquakes ($M_{LH} > 4.2$) occurring in the major part of Slovenia are between 9 and 20 km. This is the case in central and eastern Slovenia, as well as in Friuli and the southwestern part of the region under investigation. Deeper earthquakes with depths exceeding 25 km are probably connected with major faults reaching the basement, presumably separating the larger geotectonic units in the region. They occur in Friuli (Italy) and in the Rijeka (Croatia) region. Hypocentres are also deeper in the region between eastern Slovenia and Croatia.

Only recently, the increased number of instruments and tools for data-processing has allowed detailed seismicity-mapping based on instrumental data. The distribution of recent seismicity relocated using instrumental data for the period from 1989 to 1996 is presented on Figure 5. Some distinct lines of hypocentral concentration also resulted from the simultaneous inversion for hypocentres and a three-dimensional crustal velocity model (Figs. 3 and 4 in MICHELINI *et al.*, 1998). The most pronounced are the delineations along the Raša and the Idrija faults in the NW-SE direction. The hypocentral depth distribution shows a similar pattern, however earthquakes are, on average, deeper than those estimated from macroseismic data for older earthquakes.

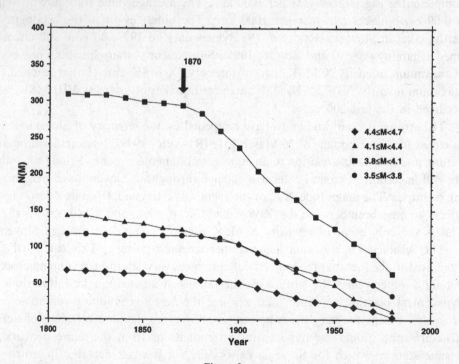

Figure 3

Cumulative number of earthquakes vs. time for several values of magnitudes. Under the assumption that the seismicity is stationary, one can judge that the catalogue is complete for earthquakes of magnitude 3.8 and larger for the period after 1870.

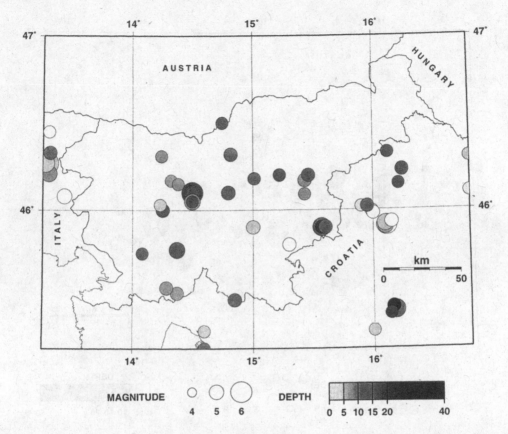

Figure 4

Hypocentres of earthquakes of $M_{LH} \geq 4.5$, in the period 1871–1995. Epicentres and focal depths are determined from macroseismic data. The error of the macroseismic epicentre determination is estimated to be about 0.1 degree.

Fault plane solution determinations for earthquakes in Slovenia and Croatia were taken from the literature (HERAK *et al.*, 1995; RENNER and SLEJKO, 1994; CARULLI *et al.*, 1990; PONDRELLI *et al.*, 1998) and the unpublished reports (ŽIVČIĆ, 1994), and were supplemeted with additional determinations using the polarities and amplitudes of the first arrival of longitudinal and transversal waves (SNOKE *et al.*, 1984) recorded on local networks. In total, a selection of well constrained solutions for thirty earthquakes was collected (Fig. 6). Magnitudes of earthquakes range from 2.2 to 5.6. Although strike-slip and thrust-type dominate, there are also a few earthquakes with normal-type faulting. Probably due to the relatively small magnitude of the earthquakes used, most of the earthquake mechanisms seem to be controlled by local conditions rather than by those prevailing on a regional scale.

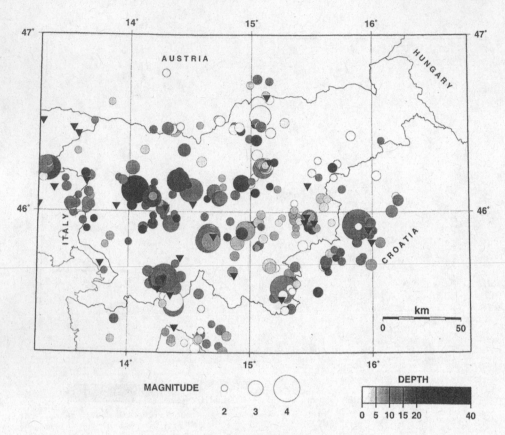

Figure 5

Hypocentres of earthquakes of $M_L \geq 2.0$, in the period 1989–1996. Epicentres and focal depths are determined from instrumental records using HYPOELLIPSE (LAHR, 1993) and 1-D velocity model. Only epicentres determined with the standard deviation less than 5 km are shown. Stations used for location are given as inverted triangles. Not all stations were active during the entire period.

Geodynamics

From the fault plane solutions (FPS) it is evident that the governing stress in the region runs approximately in a N-S direction (Fig. 6). We weighted individual fault plane solution according to its reliability, taking into consideration the proportion of weights of the readings in error, the number of inconsistent readings and the degree to which the solution is constrained by the data into three classes. The solutions that were not constrained to better than 30 degrees were discarded. We estimate that on average nodal plane orientation is constrained to within 20 degrees. Under the assumption of uniform stress throughout the region, the principal stress σ_1 is determined using the method of GEPHART and FORSYTH (1984). Its azimuth is 6° and dip 8°. Dip of the least principal stress σ_3 of 5° is consistent with regional strike slip regime. In addition, the horizontal stress field in

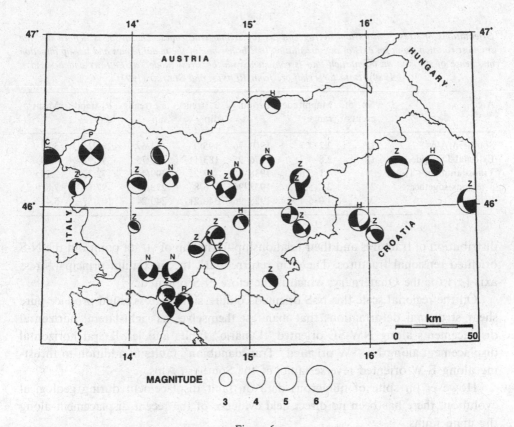

Figure 6

Fault plane solutions of earthquakes. Letters above the beachballs correspond to the reference used:
C—CARULLI *et al.*, 1990; H—HERAK *et al.*, 1995; R—RENNER and SLEJKO, 1994; ŽIVČIĆ, 1994;
P—PONDRELLI *et al.*, 1998; N—this study. For Friuli (Italy) region only the main shock of the 1976
sequence is shown.

the wider region, as determined from focal mechanisms (UDIAS and BUFORN, 1993)
and by modelling (GRÜNTHAL and STROMEYER, 1986), has an approximately
N-S-oriented maximum principal stress. However, the largest smaller misfit of
individual nodal planes is 39 degrees, another two being larger than 25 and ten
larger than 10 degrees, suggesting that the stress is not uniform within the volume
considered. Inverting for the stress in three separate regions we achieved better
results: the largest smaller misfit of individual nodal planes is 26 degrees, another
four being larger than 10 and the rest are less than 10 degrees. These results suggest
that the stress regime within different seismogenic areas is uniform. The results of
stress inversion are given in Table 1.

Within this study, we have also performed field structural measurements in
Quaternary river terraces along the Sava river northwest of Ljubljana. These are
slightly dipping towards the north, and they are systematically fractured. The

Table 1

Orientations of stress tensors and average misfits for seismogenic zones. The misfit is defined as "the smallest rotation about an axis of any orientation that brings one of the nodal planes and its slip direction and sense of slip into an orientation that is consistent with the stress model" (GEPHART and FORSYTH, 1984). Data for Friuli are from RENNER and SLEJKO (1997)

Area	No. of events	Magnitude range	Period	σ_1 trend/dip	σ_2 trend/dip	σ_3 trend/dip	Misfit
Southern Alps	12	2.2–4.7	1963–96	195/9	84/67	288/21	5.47
External Dinarides	10	2.8–4.7	1976–95	183/11	323/75	91/9	4.40
Transdanubian + Tisza	5	4.1–5.6	1938–96	67/22	309/49	172/33	2.74
3 regions together	27	2.2–5.6	1938–96	6/8	223/81	97/5	7.87
Friuli	21	4.0–6.4	1928–96	340/21	240/24	107/57	5.4

distribution of fractures and their relationship to the dip of strata point out the N-S oriented tensional fractures. These, in return, mark the maximum principal stress axis (σ_1) for the Quaternary, which is therefore N-S oriented.

On the regional scale this N-S oriented compressional stress could produce pure shear structural deformations that manifest themselves as right-lateral horizontal displacements along NW-SE oriented "Dinaric" faults and left-lateral horizontal displacement along NE-SW oriented "Transdanubian" faults in addition to thrusting along E-W oriented reverse faults of the Southern Alps.

However, in spite of numerous data on fault displacement during geological evolution, there has been no direct field evidence of the recent displacement along the main faults.

The data on vertical movements obtained by geodetic measurements in the last 100 years have been processed by KOLER (1990). The data show a relative uplift of the area in central and northern Slovenia (Southern Alps) of up to 8 mm/year, as compared to the terrain in southern Slovenia (Istria and the southwestern part of the Dinarides), and 7 mm/year relative to the Pannonian basin, which is significantly above mean error of the estimate of 0.3 mm/year. The density of measurements (benchmarks) is, unfortunately, not high and does not enable the more precise determination of possible tectonic structures and their activity. Differential movements on a more local scale were measured in the Krško basin (in southeastern Slovenia), and the rate less than 1 ± 0.2 mm/year was obtained (KOLER and BREZNIKAR, 1998). This tallies well with the values calculated from the differences in the elevations of various Neogene stratigraphic units in the same area (POLJAK and ŽIVČIĆ, 1994, 1995).

The first attempts to measure horizontal movements along a number of faults in Slovenia were made in the 1970s. Six geodetic networks were installed: three in the Ljubljana area, then in the Karavanke Mts., along the Idrija fault and in the Krško basin. With the exception of the Krško basin, no displacements have been proven, either due to the short timespan or the inadequate position of nodal points in the network (KOGOJ, 1997). The geodetic measurements of horizontal movements, using GPS across Slovenia, have been established only recently, and have not yet yielded accurate results. However, the local network in the Krško basin suggests displacements of less than 1 mm/year. This rate agrees with the values, calculated from the total offset since the Neogene, along several faults in the same area (POLJAK and ŽIVČIĆ, 1994, 1995).

Seismogenic Areas

The data presented on tectonics and seismicity suggest that the territory of Slovenia, along with its neighbouring regions, has several characteristics that are recognisable and can be distinguished from one another.

There is general agreement between a tectonic pattern and its geological dynamics with the type of seismicity of a particular area. These areas generally coincide with the main geotectonic units previously described (Fig. 7). However, some parts express differences, which must be taken into consideration when explaining the seismotectonics of a given area.

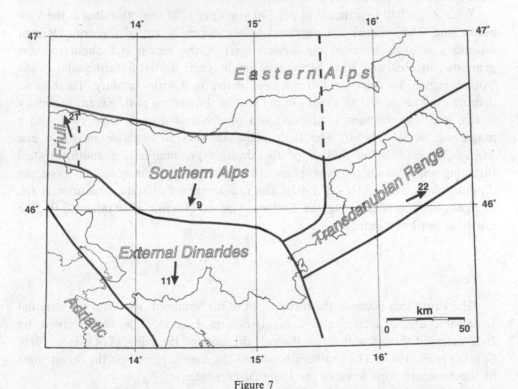

Figure 7

Seismogenic areas of Slovenia and adjacent regions. Arrows show the direction of maximum principal stress as determined from fault plane solutions with numbers for the dip.

1. Eastern Alps

The Eastern Alps lie north of the Periadriatic line. According to the structural geometry and dynamics, they could be divided into two main parts: western and eastern.

The western part of the Eastern Alps is characterised by the northward thrusting of the fore-thrusting type, considering the Periadriatic line as the consumed subduction zone between the European plate and the Adriatic microplate (PFIFFNER, 1992). After the Cretaceous-Paleogene subduction, the Periadriatic line developed the character of a strike-slip fault with a dextral sense of horizontal displacement (BEMMELEN, 1970).

The eastern part of the Eastern Alps is mostly covered by the Tertiary sediments of the Pannonian basin. The Ljutomer fault in Slovenia and the Balaton fault in Hungary are considered to be the southeastern boundary of the Eastern Alps with the Southern Alps (RAKOVEC, 1956; PLENIČAR, 1969; HORVÁTH, 1993). At the same time, these two faults, in addition to a number of others (i.e., the Raba and the Balaton faults), represent large strike-slip faults with a sinistral sense of displacement (PLENIČAR, 1969; RUMPLER and HORVÁTH, 1988). The northward 'fore-thrusting' is not clearly detectable.

With $A_{3.8} = 0.014$ earthquakes per 100 years per 1000 km^2, this area is the least active one. The seismicity is restricted to the upper 20 km of the crust. Recent seismicity is low. However, the western part of the region is thought to have generated at least one large earthquake in the past: the 1690 earthquake in the Villach region. Its epicentre has not been determined with certainty. There is no definite evidence of strong shocks in the past in the eastern part. Recent seismicity is also low. The strongest earthquake is a poorly-located event from 1943 with a magnitude of $M_{LH} = 5.0$. The fault plane solution is available only for one $M_{WA} = 3.6$ earthquake, and is of the thrust type, manifesting southwestward thrusting on a gently dipping plane, or northeastward thrusting on a steeply dipping plane (HERAK *et al.*, 1995). The measurements of some structures in the Neogene sediments covering the Eastern Alps in Slovenia (POLJAK, 1984) gave north to northwest oriented σ_1.

2. Friuli

The Friuli area occupies the thrust front of the Southern Alps onto the External Dinarides, and is tectonically a highly-deformed area. It is characterised by E-W-oriented thrusts with a southward direction of thrusting (TOLLMAN, 1986; SLEJKO *et al.*, 1987). The southern border of the zone is placed on the thrust front of the Southern Alps towards the External Dinarides.

Geophysical data show that the Mesozoic basement south of the Southern Alps thrust is built of Dinaric (NW-SE) thrusts with a southwestward displacement

(CATI et al., 1989). These thrusts change their position west of the river Tagliamento. There they are NE-SW-oriented with a southeastward thrusting direction. The junction of these two sets of thrusts is the trough along the northern Tagliamento river valley, bordered by NNE-SSW- and NNW-SSE-oriented faults that express a horizontal displacement. These faults continue northwards into the Southern Alps.

This is the most active zone in the area under investigation ($A_{3.8} = 0.67$). There is substantial evidence of continuous seismic activity throughout history, with earthquakes with a maximum intensity IX MCS. Some of the strongest earthquakes in the entire region, whose epicentral location is rather uncertain, are presumed to have occurred inside or at the northern and eastern margins of this region (the so-called 1348 and 1690 Villach events, and the 1511 Idrija events). The most recent seismic sequence occurred in 1976, with the strongest earthquake of $M_{LH} = 6.2$. Most fault plane solutions indicate thrusting, while a smaller number, restricted mostly to the Tagliamento trough is of the strike-slip type. Hypocentral depths do not exceed 15 km. The maximum stress direction in this zone is 340° dipping 21°, while the least stress direction plunges 57° (RENNER and SLEJKO, 1997).

3. Southern Alps

In Slovenia, the Southern Alps encompass the Southern Karavanke, the Julian and the Kamnik-Savinja Alps, and the area of the central Slovenia.

The Southern Alps are characterised by regional thrusts from the north to the south, among which the thrusts of the Julian and Savinja-Kamnik Alps are the most prominent. In addition to thrusting, a number of regional faults in a NW-SE direction can be detected. They all exhibit a dextral sense of horizontal displacement of the Neogene to Quaternary age (PLENIČAR, 1969). According to PLACER (1996b), the total displacement along the Sava fault since the Paleogene is between 65 and 70 km. This type of structural pattern can be followed to the western rim of the Pannonian basin in eastern Slovenia. These faults originate from the Paleogene External Dinaric tectonic evolution, and they are superimposed by the Neogene Southern Alps thrusts. They have been reactivated in fragments in the post-Neogene time (PLACER, 1996a).

The seismicity of the zone is moderate ($A_{3.8} = 0.13$) and rather shallow. Most earthquakes happen in the upper 15 km. The strongest shock occurred in 1895 and it damaged the city of Ljubljana. Its magnitude was estimated from intensity data as being $M_{LH} = 6.1$ ($I_0 = $ VIII–IX MSK). From twelve fault plane solutions the principal stress in the zone is determined to have a direction of 195° and dip of 9°. The gently-dipping least-stress direction (21 degrees) suggests a dominant strike-slip mechanism. Strike-slip faults and thrust faults, probably connected with southward thrusting, are common, although normal faulting is also present. The strongest shock of this century (in 1963 with $M_{LH} = 4.7$) may be related to the Sava fault. Its fault plane solution is of the strike-slip type and is consistent with tectonic movements along the Sava fault.

4. *Transdanubian Range*

This area marks the collision zone between the Tisza microplate and the European plate. Today this area is characterised, at least in Slovenia, by folds, faults and thrusts stretching in a NE-SW direction. The faults are generally assumed to be of the strike-slip type, with a left lateral sense of movement (i.e., RUMPLER and HORVÁTH, 1988), which has recently been proved in the Krško basin (POLJAK *et al.*, 1996).

This zone has been continuously active, with two poorly-determined events with intensity IX MSK in 1459 and 1640. The strongest events ($M_{LH} = 6.1$) in 1880 and 1906 happened north of Zagreb. The activity $A_{3.8}$ is 0.18. Earthquake hypocentres are concentrated in the upper 15 km, with a few isolated events reaching depths of 30 km. The available fault plane solutions show either thrust or gravitational faulting, with practically negligible strike-slip component. From five earthquakes in the zone of the Transdanubian Range and the Tisza unit, the principal stress direction is 67 degrees, with a dip of 22 degrees.

5. *External Dinarides*

The External Dinarides (Dinarides *sensu stricto*) occupy the southern part of Slovenia. The basic structural characteristics of the External Dinarides is a dense pattern of faults in a NW-SE direction, in addition to the thrusts with the southwestward direction of thrusting.

The western part of the External Dinarides occupies the southwestern part of Slovenia to the Idrija fault. Its dominant structural characteristic is a set of Dinaric faults, as well as a series of thrusts with a southwestward displacement. The southwestern border of this zone is a series of faults and thrusts in the area between Trieste and Rijeka. This zone extends further southeastwards along the Adriatic coast. Some authors (e.g., ALJINOVIĆ *et al.*, 1984) consider it to be the main discontinuity between the External Dinarides and the Adriatic basin (Adriatic). Recent investigations (POLJAK and RIŽNAR, 1996), however, have shown that there is a continuous structural and paleogeographical transition between the Adriatic and the External Dinarides. The contact zone consists of a series of folds with minor thrusting towards the southwest.

The northwestern part of the External Dinarides is covered by Tertiary and Quaternary sediments of the Po plain in Italy. Geophysical investigations (CATI *et al.*, 1989) clearly express the continuation of the Dinaric faults of the reverse type extending to the thrust front of the Southern Alps.

The eastern part of the External Dinarides occupies the area from the Idrija fault in the west to the Žužemberk fault in the east. It is also characterised by regional faults with a dextral sense of horizontal displacement, however thrusting is of a much lesser extent. There are also regional faults in a NE-SW direction that extend from the Transdanubian Range and exhibit a sinistral sense of horizontal displacement.

The External Dinarides are characterised by moderate historic and recent seismicity ($A_{3.8} = 0.09$). In the western part, earthquake foci reach to depths of 25 km, with a few events around 30 km. Relocations of the hypocentres in the last ten years have delineated the hypocentral alignments along the Raša and Idrija faults in NW-SE direction (MICHELINI et al., 1998), most likely representing thrusts of the External Dinarides over the undeformed segment of the Adria plate (Adriatic on the Fig. 1). However, fault plane solutions of several of these earthquakes, as well as their distribution in a vertical plane, indicate that the recent seismicity of these faults is of the right-lateral strike-slip type. Few earthquakes have fault plane solutions of normal faulting. The strongest shock, with a magnitude of $M_{LH} = 5.8$, took place in 1916 to the southeast of Rijeka.

The eastern part of the zone has practically no record of historical seismicity. Recent seismicity reaches depths of 15 km, with the strongest shock of a magnitude of $M_{LH} = 5.6$. Most fault plane solutions are of a thrust type towards the south or southwest. Based on ten fault plane solutions, the principal stress in the External Dinarides has an azimuth of 183 degrees and a dip of 11 degrees, whereas the nearly horizontal least-stress axis and vertical intermediate stress axis suggest earthquake occurrence on strike-slip faults.

The Idrija fault separates the western and the eastern parts of the External Dinarides in southern Slovenia. It is vertical or subvertical, with a dip towards the northeast. Horizontal and vertical movements have been documented (MLAKAR, 1964; PLACER, 1981). The fault has been mapped to a depth of 1 km, and a deep seismic-sounding profile (Joksović and ANDRIĆ, 1983) has identified the probable continuation of the fault down to between 5 and 6 km. The strongest earthquake in the whole area under study (the 'Idrija' earthquake of 1511 with an estimated magnitude of 6.8) is usually related to this fault (RIBARIČ, 1979). However, recent seismicity in the vicinity of the fault is rather low, with only a few events reaching intensity VI MSK. The strongest post-1511 event happened in 1926 at the southeastern end of the mapped fault, and it had a magnitude of 5.6.

Conclusions

The integration of tectonic and seismological data, in addition to geodetic and stress measurements data illustrates that the area under investigation is tectonically and seismically a highly complex one. The recent structural pattern is a cumulative result of Tethyan evolution, where recent dynamics is determined by the closure of the Tethys and the collision of several lithospheric units.

The largest part of Slovenia, together with northern Italy, lies at the contact between the European plate and the Adriatic microplate, whereas the Southern Alps and the Dinarides represent the deformed rim of the Adriatic plate. The stress regime here is compressional with the N-S oriented σ_1. The resulting deformation is mostly southward thrusting in the Southern Alps, as well as strike-slip and transpression displacement along NW-SE trending faults of the External Dinarides.

The eastern part of Slovenia, together with neighbouring areas of Hungary and Croatia, lies at the contact between the European plate and the Tisza microplate. This contact is marked by a broad belt of the Southern Alps—Internal Dinarides rocks that correspond to the so-called Transdanubian Range of Central Hungary. According to seismological data, the σ_1 axis is oriented here towards the E-NE, which is in general accordance with the entire Pannonian basin (GRÜNTHAL and STROMEYER, 1986). However, the dominant structural deformation, at least in Slovenia (PLENIČAR, 1969), is reported to be left-lateral horizontal displacement along the NE-SW oriented faults. Therefore, the maximum principal stress axis in the southwestern part of the Pannonian basin should be more N-S oriented.

The northward anti-clockwise movement of the Adriatic microplate, and the northward clockwise movement of the Tisza microplate in relation to the relatively stable European plate, resulted in a compressional stress field with a generally N-S-oriented σ_1 in the Southern Alps and External Dinarides, N-NW in Friuli and E-NE in the Pannonian basin. This stress field caused the southward thrusting of the Southern Alps, dextral displacement along the NW-SE-oriented faults and the transpression of the same orientation, and the sinistral displacement of the Transdanubian Range NE-SW-oriented faults, with possible transpression in the same direction. Thus the NE-SW oriented strike-slip faults of the Pannonian basin, together with the NW-SE oriented strike-slip faults of the External Dinarides, form a pair-set of shear structures on a regional scale.

According to generally accepted ideas, noted lithospheric units (Adria, Europe and Tisza) have been amalgamated together during Tertiary time. The seismicity is not concentrated along the primary plate boundary but is rather spread in a broad zone along their deformed rims. This is in accordance with moderate seismicity, with maximum earthquake magnitudes up to 7, and moderate deformation rates.

Despite the fact that the tectonics of the area under investigation is relatively well understood, direct evidence of recent deformation is less obvious. This is primarily related to visible or detectable displacements along faults. To date no such displacement has been found in recent sediments, and no surface rupture accompanying seismic event has been detected. Therefore, it is not possible to relate a particular seismic event to a certain tectonic structure and to make detailed seismotectonic models of this area.

The area has been divided into five seismogenic areas: the Eastern Alps, the Southern Alps, with the Friuli region as a separate unit, the External Dinarides, and the Transdanubian Range. Their structural patterns display similar geological evolution and recent activity, although differ from unit to unit. The seismic characteristics are also recognisable in a general sense with minor differences, which result in local stress fields, structural deformations and seismicity. This zoning may represent the framework for further detailed qualitative and quantitative seismotectonic analysis of this area.

Acknowledgements

This work was supported by the EC Copernicus contract CIPA-CT94-0238. We have used GMT public domain graphics software for figures (WESSEL and SMITH, 1991) G. Renner and D. Slejko kindly provided a pre-print of their paper. We are grateful for the comments provided by H. Bungum, an anonymous reviewer and the editors that considerably helped improving the paper.

REFERENCES

ALJINOVIĆ, B., BLAŠKOVIĆ, I., CVIJANOVIĆ, D., PRELOGOVIĆ, SKOKO, D., and BRDAREVIĆ, N. (1984) *Correlation of Geophysical, Geological and Seismological Data in the Coastal Part of Yugoslavia*, Boll. Ocean. Teor. Appl. *2*, 77–90.

BEMMELEN, R. W. VAN (1970), *Tektonische Probleme- der Östlichen Südalpen*, Geologija *13*, 133–158, Ljubljana.

BUSER, S. (1989), *Development of the Dinaric and Julian Carbonate Platforms, and of the Intermediate Slovene Basin (NW Yugoslavia)*, Mem. Soc. Geol. Ital. *40*, 313–320.

BUSER, S., and DRAKSLER, V. (1990), *Geološka karta Slovenije 1:500000*. Mladinska knjiga, Ljubljana.

CARULLI, G. B., NICOLICH, R., REBEZ, A., and SLEJKO, D. (1990), *Seismotectonics of the Northwest External Dinarides*, Tectonophysics *179*, 11–25.

CATI, A., FISCHERA, R., and CAPPELLI, U. (1989), *Northeastern Italy, Integrated Processing of Geophysical and Geologic Data*, Mem. Soc. Geol. Ital. *40*, 273–288.

CECIĆ, I., and ŽIVČIĆ, M. (1996), *The Oldest Earthquake in the Slovene Catalogue*, The XXV General Assembly of the ESC, Reykjavik, Iceland, Abstracts, p. 142.

CVIJANOVIĆ, D. (1981), *Seizmičnost produčja SR Hrvatske*, Disertacija, Sveučilište u Zagrebu, Zagreb.

DEL BEN, A., FINETTI, I., REBEZ, A., and SLEJKO, D. (1991), *Seismicity and Seismotectonics at the Alps–Dinarides Contact*, Boll. Geof. Teor. Appl. *33*, 155–176.

GEPHART, J. W., and FORSYTH, D. (1984), *An Improved Method for Determining the Regional Stress Tensor Using Earthquake Focal Mechanism Data: Application to the San Fernando Earthquake Sequence*, J. Geoph. Res. *89*, 9305–9320.

GRÜNTHAL, G., and STROMEYER, D. (1986), *Stress Pattern in Central Europe and Adjacent Areas*, Gerlands Beitr. Geophys. *95* (5), 443–452.

HERAK, M., HERAK, D., and MARKUŠIĆ, S. (1995), *Fault Plane Solutions for Earthquakes (1956–1995) in Croatia and Neighbouring Regions*, Geofizika *12*, 43–56.

HORVÁTH, F. (1993), *Towards a Mechanical Model for the Formation of the Pannonian Basin*, Tectonophysics *226*, 333–357.

JOKSOVIĆ, P., and ANDRIĆ, B. (1983), *Ispitivanje građe zemljine kore metodom dubokog seizmičkog sondiranja na profilu Pula–Maribor*. Int. Publ., Geofizika, Zagreb, 13 pp.

KARNÍK, V. (1968), *Seismicity of the European Area, Part 1*, Academia, Czechoslovak Academy of Sciences, Praha.

KOBER, L. (1921), *Der Bau der Erde*, Gebrüder Borntraeger, Wien, 324 pp.

KOBER, L. (1931), *Das Alpine Europa und sein Rahmen*, Gebrüder Borntraeger, Wien, 310 pp.

KOGOJ, D. (1997), *Geodetske meritve stabilnosti tal ob tektonskih prelomih na območju Slovenije*, Zbornik SZGG *3*, 133–144.

KOLER, B. (1990), *Ugotovitev vertikalnih permikov na osnovi analize nivelmanskih mrež višjih redov na območju Slovenije*, Disertacija, FAGG, Univerza v Ljubljani, 198 pp.

KOLER, B., and BREZNIKAR, A., *Stability of area around the nuclear power station Krško*. In *Proceedings of the 1st International Conference of Engineering Surveying* (eds. Staněk and Kopáčik) (Bratislava 1998) pp. 245–250.

KUŠČER, D. (1990), *The Quaternary Valley Fills of the Sava River and Neotectonics*, Geologija 3, 299–313, Ljubljana (in Slovene with English abstract).

LAHR, J. C. (1993), *HYPOELLIPSE/Ver.2.0: A Computer Program for Determining Local Earthquake Hypocentral Parameters, Magnitude, and First Motion Pattern*, USGS Open-File report 89-116.

LEMOIN, M. (ed.), *Geological Atlas of Alpine Europe* (Elsevier, Amsterdam 1978) 584 pp.

MICHELINI, A., ŽIVČIĆ, M., SUHADOLC, P. (1998), *Simultaneous Inversion for Velocity Structure and Hypocenters in Slovenia*, J. Seismol. *2*, 257–265.

MLAKAR, I. (1964), *Vloga postrudne tektonike pri iskanju novih orudenih con na območju Idrije*, Rudarsko-metalurški zbornik *1*, 19–25.

PFIFFNER, A., *Alpine orogeny*. In *A Continent Revealed, The European Geotraverse* (eds. Blundell, D., Free, R., Mueller, St.) (Cambridge Univ. Press, Cambridge 1992) pp. 180–190.

PLACER, L. (1981), *Tektonski razvoj idrijskega rudišča*, Geologija *25/1*, 7–94.

PLACER, L., *Tectonic structure of southwest Slovenia*. In *The Role of Impact Processes in the Geological and Biological Evolution of Planet Earth* (eds. Drobne, K., Goričan, Š., and Kotnik, B.) (Ljubljana 1996a) pp. 137–140.

PLACER, L. (1996b), *Displacement Along the Sava Fault*, Geologija *39*, 283–287.

PLENIČAR, M. (1969), *Tektonska karta Slovenije 1:200,000 (Tectonic Map of Slovenia, 1:200,000)*, Geological Survey Ljubljana.

POLJAK, M. (1984), *Neotectonic Investigations in the Pannonian Basin Based on Satellite Images*, Adv. Space Res. *4* (11), 139–146.

POLJAK, M. (1986), *The Structural Evolution of the Slovene Outer Dinarides in the Tertiary and Quaternary*, Proceed. 11. Congr. Geol. Yug., 3, 299–322, Tara (in Croatian with English abstract).

POLJAK, M., and RIŽNAR, I. (1996), *Structure of the Adriatic-Dinaric Platform Along the Sečovlje-Postojna Profile*, Geol. Croat. *49/2*, 345–346, Zagreb.

POLJAK, M., VERBIČ, T., GOSAR, A., ŽIVČIĆ, M., and RIBIČIČ, M. (1996), *Neotectonic Investigations in the Krško NPP Vicinity*, Vol. 5. Int. Publ. Geological Survey Ljubljana, 70 pp. (in Slovene with English abstract).

POLJAK, M., and ŽIVČIĆ, M. (1994), *Seismic sources*. In *Probabilistic Assessment of Seismic Hazard at Krško Nuclear Power Plant* (eds. Fajfar, P., and Lapajne, J.), Rev. 1, FAGG, Ljubljana (unpublished report).

POLJAK, M., and ŽIVČIĆ, M. (1995), *Tectonics and Seismicity of the Krško Basin*, Proceed. 1. Croat. Geol. Congr. *2*, 475–479, Zagreb.

PONDRELLI, S., MORELLI, A., and EKSTRÖM, G. (1998), *Moment Tensors and Seismotectonics of the Mediterranean Region*, The EGS XXXII General Assembly, Nice, France, 20–24 April, 1998.

RAKOVEC, I. (1956), *A Survey of the Tectonic Structure of Slovenia*, Proceed. 1. Yug. Geol. Congr., 73–83, Ljubljana (in Slovene with English abstract).

RENNER, G., and SLEJKO, D. (1994), *Some Comments on the Seismicity of the Adriatic Region*, Boll. Geof. Teor. Appl. *36*, 381–398.

RENNER, G., and SLEJKO, D. (1997), *Stress Tensor Computation from Fault Plane Solutions for the Eastern Adriatic Region*, 29th General Assembly of the IASPEI, Thessaloniki, Greece, Abstracts, p. 18.

RIBARIČ, V. (1979), *The Idrija Earthquake of March 26, 1511*, Tectonophysics *53*, 315–324.

RIBARIČ, V. (1982), *Seismicity of Slovenia—Catalogue of Earthquakes (792 A.D-1981)*, SZ SRS, Publication, Ser. A, No. 1-1, Ljubljana, 650 pp.

RUMPLER, J., and HORVÁTH, F. (1988), *Some representative seismic reflection lines from the Pannonian Basin and their structural interpretations.* In *The Pannonian Basin*, Amer. Assoc. Petrol. Geol. Mem. (eds. Royden, L. H., and Horváth, F.) *45*, 153–167.

SHEBALIN, N. V., KÁRNÍK, V., and HADŽIEVSKI, D. (1974), *UNDP/UNESCO Survey of the Seismicity of the Balkan Region, Catalogue of Earthquakes*, Skopje.

SLEJKO, D., CARRARO, F., CARULLI, G. B., CASTALDINI, D., CAVALLIN, A., DOGLIONI, C., ILICETO, V., NICOLICH, R., REBERZ, A., SEMENZA, E., ZANFERRARI, A., and ZANOLLA, C. (1987) *Modello sismotettonico dell'Italia nord-orientale.* C.N.R., G.N.D.T. Rend. 1, 82 pp.

SNOKE, J. A., MUNSEY, J. W., TEAGUE, A. G., and BOLLINGER, G. A. (1984), *A Program for Focal Mechanism Determination by the Combined Use of Polarity and SV-P Amplitude Ratio Data*, Earthquake Notes *55*, No. 3, p. 15.

TOLLMAN, A. (1986), *Tektonische Karte von Österreich, 1:500,000*, Geol. Bundesanst., Band 3, Wien.

UDIAS, A., and BUFORN, E., *Regional stresses in the Mediterranean region derived from focal mechanisms of Earthquakes.* In *Recent Evolution and Seismicity of the Mediterranean Region* (eds. Boschi, E., Mantovani, E., and Morelli, A.) (Academic Publishers, Kluwer 1993) pp. 261–268.

WESSEL, P., and SMITH, W. H. F. (1991), *Free Software Helps Map and Display Data, EOS*, Trans. Amer. Un. *72* (441), 445–446.

ŽIVČIĆ, M. (1984), *Odredivanje dubine žarišta potresa na osnovi makroseizmičkih podataka*, Geofizika *1*, 217–221.

ŽIVČIĆ, M., *Earthquake catalogue.* In *Probabilistic Assessment of Seismic Hazard at Krško Nuclear Power Plant* (eds. Fajfar, P., and Lapajne, J.) (University of Ljubljana, Dept. of Civil Engineering 1994).

ŽIVČIĆ, M., and CECIĆ, I. (1998), *Revised magnitudes of historical earthquakes in Slovenia*, The EGS XXXII General Assembly, Nice, France, 20–24 April, 1998.

ŽIVČIĆ, M., SUHADOLC, P., and VACCARI, F. (2000), *Seismic Zoning of Slovenia Based on Deterministic Hazard Computations*, Pure appl. geophys. *157*, 171–184.

ŽLEBNIK, L. (1971), *Pleistocene Deposits of the Kranj, Sava and Ljubljana Fields*, Geologija *14*, 5–51, Ljubljana (in Slovene with English abstract).

(Received March 27, 1998, revised November 27, 1998, accepted May 11, 1999)

To access this journal online:
http://www.birkhauser.ch

Pure appl. geophys. 157 (2000) 57–77
0033–4553/00/020057–21 $ 1.50 + 0.20/0

Pure and Applied Geophysics

Characterization of Seismogenic Zones of Romania

M. RADULIAN,[1] N. MÂNDRESCU,[1] G. F. PANZA,[2,3] E. POPESCU[1] and A. UTALE[1]

Abstract—Although the time and magnitude range covered by available seismological data is limited, several significant regional trends are outlined in the seismogenic zones of Romania. Vrancea region, which is by far the most seismically active area, has a persistent rate of occurrence of intermediate-depth earthquakes, clustered in a very confined focal volume, and a clear compressive stress regime. The deformation field, as deduced from the available fault plane solutions, is drastically reduced in the crust, where the maximum magnitude is below 6.5 (except Shabla zone, in Bulgaria). The system of major faults developed in a NW–SE direction in the Carpathians foredeep area is certainly linked to the subduction process in Vrancea, although they seem not to play a significantly active role, as could be expected for an active subduction process. The existing data indicate an extensional deformation regime over the foredeep area and Southern Carpathians, while a predominant compressive regime is outlined at the contact between the eastern margin of the Pannonian Depression and Carpathians orogen, in agreement with the bending tendency of the maximum horizontal compression orientation of the crustal stress field from NE–SW, in western and central Europe, to E–W, in the intra-Carpathian region (GRÜNTHAL and STROMEYER, 1992).

Key words: Seismogenic zone, stress field, Carpathians area, Vrancea subduction.

Introduction

The seismogenic zones are areas of grouped seismicity in which the seismic activity and stress field orientation are assumed to be relatively uniform. The identification of long-term characteristics of the earthquake generation process in each seismogenic zone is of great significance for the seismic hazard assessment. Certainly this implies the availability of data covering the time scale of the tectonic processes, which is not the case for the earthquake catalogues presently available. However, the main trends identified in the reported catalogues could be representative of the potential earthquakes, at least in some cases, as shown by PANZA *et al.* (1997) for the Vrancea seismic region.

The present study is an extension of the work by RADULIAN *et al.* (1996), in which the seismogenic zones on the Romanian territory are identified on the basis

---

[1] National Institute for Earth Physics, Bucharest, Romania.
[2] The Abdus Salam International Centre for Theoretical Physics, SAND Group, Trieste, Italy.
[3] Dipartimento di Scienze della Terra, Universita' di Trieste, Italy.

of tectonics and seismicity information, and are correlated with the main tectonic units, active faults and epicenter distribution.

The earlier schemes proposed by RADU *et al.* (1980) and CONSTANTINESCU and MÂRZA (1980), to divide the Romanian territory into seismogenic areas, follow simply the geographical distribution of the seismic activity and the configuration of the administrative provinces, bearing no connection with the characteristics of regional tectonics. The seismic zones we defined, on the basis of tectonic units and seismicity, are meant to represent, even schematically, the surface projection of the system of active faults.

The continental tectonics is not clear and simple to interpret in general, and the presence of very thick Neogene and Quaternary deposits over a large part of the Romanian territory makes particularly difficult the identification of the active faults and a quantitative specification of their kinematics. Therefore we must be less ambitious and restrict our approach to a relatively descriptive level. Despite these limitations, our work may be justified, as the spatial distribution of earthquakes in any region of Romania could be considered as an areal source (diffused zone), i.e., an area where faults are too numerous, randomly oriented and difficult to be defined individually. This is valid also for the few areas where major faults are identified (in the Carpathians foredeep region).

The geographical distribution of the seismogenic zones and the distribution of the epicenters of the crustal earthquakes are given in Figure 1. The Shabla zone, situated in the northeastern part of Bulgaria close to the border with Romania, is also considered. To the east, the earthquakes are related to the subduction process at the Carpathians arc bend (Vrancea region); to the west, they follow the contact between the Pannonian Depression and the Carpathians orogen. The eastern segment of the Carpathians in Romania is practically aseismic, except the southern extremity (Vrancea region). The western segment (Apuseni Mountains) is aseismic as well. The southern Carpathians are considerably more seismically active, especially in the eastern (zone FC) and western (zone DA) extremities. The platform regions are stable, except the small strip crossing the Carpathians foredeep area on a SW–NE direction, in front of Vrancea region. Several active faults are identified here following the same SE–NW orientation (Intramoesian, Perceneaga-Camena,

---

Figure 1

Geographical distribution of the seismogenic zones and crustal seismicity. VR: Vrancea; EV: East Vrancea; BD: Bârlad Depression; PD: Predobrogean Depression; IM: Intramoesian Fault; SH: Shabla; FC: Făgăraş-Câmpulung; DA: Danubian Zone; BA: Banat; CM: Crişana-Maramureş; TD: Transylvanian Depression. Solid lines: border limits of the seismogenic zones; dotted lines: border of tectonic units: dashed lines: major active faults. IMF: Intramoesian Fault; PCF: Peceneaga-Camena Fault; SGF: Sfântul Gheorghe Fault; TF: Trotuş Fault. Inset: Tectonic sketch of Romania: 1 = Carpathian orogenic belt; 2 = Focşani-Odobeşti Depression; 3 = Fore-Carpathian zone; 4 = plate-subplate boundary; 5 = subplate–subplate boundary; 6 = strike-slip fault; 7 = active subduction; 8 = Neogene "frozen" subduction; 9 = intra-plate crustal fracture; 10 = (a) crustal and (b) subcrustal earthquake epicenters.

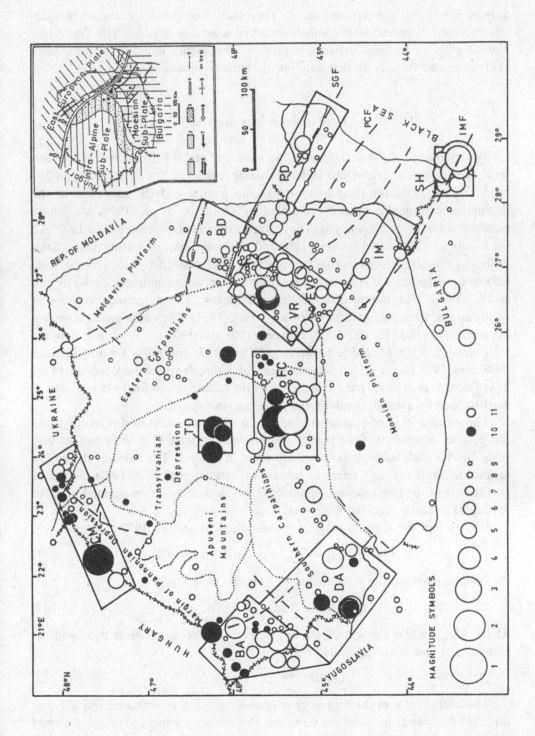

Sfântul Gheorghe and Trotuş faults). They mark the contact between different tectonic units, where a relative enhancement of seismicity appears. The Transylvanian Depression is almost aseismic at present. The small isolated seismogenic zone (TD) delimited there is defined only on the basis of historical earthquakes.

Data and Analysis Method

To determine the characteristics of the seismicity and deformation field for each seismogenic zone, a revised and updated catalogue is used. The catalogue compiled by MUSSON (1996) for the Circum-Pannonian region and the catalogue of the Romanian earthquakes recently compiled by ONCESCU *et al.* (1999) are jointly considered for seismicity analysis. They cover a time interval from 984 to 1997. All the events which occurred between 1980 and 1997 were relocated using digital data. Different magnitude scales usually considered were converted into a single scale (moment magnitude M_w) in order to homogenize the magnitudes of large and small, crustal and intermediate-depth earthquakes. The magnitude conversion relationships are synthesized in ONCESCU *et al.* (1999). The catalogue is complete between 1411–1800 for $M_w \geq 7.0$, between 1800 and 1900 for $M_w \geq 6.5$, between 1901 and 1935 for $M_w \geq 5.5$, between 1936 and 1977 for $M_w \geq 4.5$ and between 1978 and 1997 for $M_w \geq 3.0$. A catalogue of fault plane solutions compiled by MÂNDRESCU *et al.* (1997) and completed by the same authors for 1995 and 1996 is used to identify average trends in the deformation pattern.

The number of earthquakes with $M_w > 5.0$ which occurred in this century, and the number of available fault plane solutions with at least 15 P-wave first motion polarities for each seismogenic zone, are given in Table 1. The value of the largest seismic moment ($M_{0,max}$) refers to the entire time interval (984–1997), while the coefficients of the frequency-magnitude relation (a and b) are determined only for earthquakes which occurred after 1900.

If we adopt the moment-magnitude relation (KANAMORI, 1977):

$$\log M_0[\text{Nm}] = c\, M_w + d \quad (c = 1.5, d = 9.1) \tag{1}$$

the seismic moment rate is estimated by (MOLNAR, 1979):

$$M_0 = A/(1 - B)\, M_{0,max}^{1-B} \quad [\text{Nm/year}] \tag{2}$$

where $M_{0,max}$ is the moment of the largest event in the seismogenic zone and the constants A and B are calculated from:

$$A = 10^{(a + bd/c)}, \qquad B = b/c. \tag{3}$$

The coefficient a of the frequency-magnitude relation is normalized to a 1-year time interval, and, in order to compare the seismic moment rate for different seismogenic zones, to a unit epicentral area of 5000 km^2.

Table 1

Romanian seismogenic zones and their characteristics

Seismogenic zone	Number of events with $M_{w} \geq 5$*	$M_{w.max}$	a**	b	Seismic moment rate (Nm/year)	Number of fault plane solutions	Stress regime
Vrancea	179	7.9	4.77 ± 0.24	0.89 ± 0.04	1.2×10^{19}	28***	compression
East Vrancea	7	5.6	2.67 ± 0.73	0.86 ± 0.16	5.3×10^{15}	29	transition from compression to extension
Bârlad Depression	3	5.6	2.63 ± 0.47	0.83 ± 0.10	6.8×10^{15}	7	extension
Predobrogean Depression	2	5.2	2.40 ± 0.94	0.85 ± 0.21	1.8×10^{15}	4	extension
Intramoesian Fault	1	5.4	–	–	–	4	extension
Shabla	8	7.2	–	–	–	–	–
Făgăraş-Câmpulung	4	6.5	1.50 ± 0.47	0.65 ± 0.09	2.4×10^{16}	8	extension
Danubian	2	5.6	2.74 ± 1.01	0.97 ± 0.21	1.8×10^{15}	3	extension
Banat	8	5.6	2.60 ± 0.21	0.77 ± 0.04	1.3×10^{16}	30	compression
Crişana-Maramureş	1	6.2	3.19 ± 1.24	1.08 ± 0.28	2.8×10^{15}	–	compression
Transylvanian Depression	–	5.9	–	–	–	–	–

* Occurred this century; ** Normalized to 1 year and 5,000 km^2 epicentral area; *** Only earthquakes with $M_{w} \geq 5$.

Vrancea subcrustal Zone (VR)

The Vrancea region is a particularly complex seismic region of continental convergence characterized by at least three tectonic units in contact (inset of Fig. 1): the East European plate, Intra-Alpine and Moesian subplates (CONSTANTINESCU *et al.*, 1976). The strongest seismic activity of Romania concentrates at intermediate-depths (60–200 km) in an old, almost vertical subducting slab. The observation of one to six events with $M_{w} > 7$ per century in such an extremely confined focal volume implies a high level of active deformation ($\sim 3.5 \times 10^{-7}$ yr^{-1}), which is not clearly seen in the deformation of the crust. Even if we admit the hypothesis of a decoupling of the slab from the overlying crust (FUCHS *et al.*, 1979), how the deformation in the subcrustal range reflects in that in the crustal range remains an open question. A possible way to explain this is a model of dehydration-induced faulting as a plausible source for the Vrancea intermediate-depth earthquakes (GREEN, 1994; ISMAIL-ZADEH *et al.*, 1996).

The largest events which occurred in the Vrancea region are listed in Table 2. The epicenter distribution of the earthquakes with $M_{w} \geq 5.0$ which occurred after 1900 is shown in Figure 2. The focal mechanisms of the last major earthquakes (November 10, 1940; March 4, 1977; August 30, 1986; May 30, 1990 and May 31, 1990) are also represented. Two cross sections of the depth distribution of the foci of the Vrancea earthquakes, one parallel and the other perpendicular to the

Table 2

Vrancea intermediate-depth major events ($M_w > 6$) (ONCESCU et al., 1999)

Year	Month	Day	Time	Lat. (N°)	Lon. (E°)	Depth (km)	M_w
984	0	0	00:00	45.70	26.60	150.	7.1
1022	5	12	00:00	45.70	26.60	150.	6.5
1038	8	15	00:00	45.70	26.60	150.	7.3
1091	0	0	00:00	45.70	26.60	150.	7.1
1107	2	12	03:00	45.70	26.60	150.	7.1
1122	10	0	00:00	45.70	26.60	150.	6.2
1126	8	8	00:00	45.70	26.60	150.	7.1
1170	4	1	00:00	45.70	26.60	150.	7.3
1196	2	13	07:00	45.70	26.60	150.	7.5
1230	5	10	07:00	45.70	26.60	150.	7.3
1258	2	7	13:00	45.70	26.60	150.	7.1
1327	0	0	00:00	45.70	26.60	150.	7.3
1446	10	10	04:00	45.70	26.60	150.	7.5
1471	8	29	10:00	45.70	26.60	110.	7.5
1473	8	29	00:00	45.70	26.60	150.	7.3
1516	11	24	12:00	45.70	26.60	150.	7.5
1523	6	9	00:00	45.70	26.60	130.	6.5
1543	0	0	00:00	45.70	26.60	150.	7.1
1545	7	19	08:00	45.70	26.60	110.	7.1
1552	8	21	02:00	45.70	26.60	130.	6.5
1571	5	10	00:00	45.70	26.60	150.	7.1
1578	4	1	00:00	45.70	26.60	130.	6.5
1590	4	30	00:00	45.70	26.60	100.	7.3
1595	4	21	10:00	45.70	26.60	150.	7.1
1598	11	22	02:00	45.70	26.60	120.	6.5
1599	3	4	00:00	45.70	26.60	100.	6.1
1604	5	3	02:00	45.70	26.60	130.	6.8
1605	12	24	15:00	45.70	26.60	150.	7.1
1606	1	13	01:00	45.70	26.60	120.	6.8
1620	11	8	13:00	45.70	26.60	150.	7.5
1637	2	1	01:00	45.70	26.60	130.	7.1
1650	4	19	00:00	45.70	26.60	100.	6.5
1666	2	0	00:00	45.70	26.60	150.	6.1
1679	8	9	01:00	45.70	26.60	110.	7.5
1681	8	19	00:00	45.70	26.60	150.	7.1
1701	6	12	00:00	45.70	26.60	150.	7.1
1711	10	11	00:00	45.70	26.60	120.	6.5
1730	4	6	04:00	45.70	26.60	100.	6.1
1738	6	11	10:00	45.70	26.60	130.	7.7
1740	4	5	18:00	45.70	26.60	150.	7.3
1778	1	18	00:00	45.50	26.60	130.	6.5
1787	1	18	00:00	45.70	26.60	120.	6.5
1790	4	6	19:29	45.70	26.60	150.	7.1
1802	10	26	10:55	45.70	26.60	150.	7.9
1812	3	5	12:30	45.70	26.60	130.	6.5
1821	2	10	00:30	45.70	26.60	150.	6.6
1821	9	29	00:00	45.70	26.60	150.	6.1
1821	11	17	13:45	45.70	26.60	130.	6.5
1829	11	26	01:40	45.80	26.60	150.	7.3

Table 2 (*Continued*)

Year	Month	Day	Time	Lat. (N°)	Lon. (E°)	Depth (km)	M_w
1831	8	3	00:00	45.70	26.60	100.	6.1
1835	4	21	20:30	45.70	26.60	130.	6.5
1838	1	23	18:45	45.70	26.60	150.	7.5
1844	3	6	19:10	45.70	26.60	110.	6.0
1848	1	1	00:00	45.70	26.60	130.	6.5
1854	10	28	12:15	45.70	26.60	150.	6.5
1862	10	16	01:10	45.70	26.60	130.	6.5
1868	11	13	07:45	45.70	26.60	150.	6.8
1868	11	27	20:30	45.70	26.60	135.	6.5
1880	12	25	14:30	45.70	26.60	130.	6.8
1888	8	19	04:56	45.70	26.60	100.	6.5
1893	5	1	17:18	45.70	26.60	120.	6.2
1893	8	17	14:45	45.70	26.60	100.	7.1
1893	9	10	03:40	45.70	26.60	99.	6.5
1894	3	4	06:35	45.70	26.60	130.	6.5
1894	8	31	12:20	45.70	26.60	130.	7.1
1896	3	11	23:00	45.70	26.60	150.	6.6
1896	11	24	18:50	45.70	26.60	100.	6.1
1903	9	13	08:02	45.70	26.60	70.	6.3
1904	2	6	02:49	45.70	26.60	75.	6.6
1908	10	6	21:40	45.50	26.50	125.	7.1
1912	5	25	18:01	45.70	27.20	90.	6.7
1912	5	25	20:15	45.70	27.20	100.	6.1
1919	4	18	06:20	45.70	26.80	100.	6.1
1919	8	9	14:38	45.70	26.60	120.	6.0
1925	12	25	02:37	45.70	26.60	130.	6.1
1928	3	30	09:38	45.90	26.50	120.	6.0
1929	5	20	12:17	45.80	26.50	100.	6.0
1929	11	1	06:57	45.90	26.50	160.	6.1
1932	5	27	10:42	45.70	26.60	120.	6.0
1934	2	2	19:59	45.20	26.20	140.	6.0
1934	3	29	20:06	45.80	26.50	90.	6.6
1935	7	13	00:03	45.30	26.60	140.	6.0
1935	9	5	06:00	45.80	26.70	130.	6.0
1936	5	17	17:38	45.30	26.30	140.	6.0
1938	7	13	20:15	45.90	26.70	120.	6.0
1939	9	5	06:02	45.90	26.70	120.	6.2
1940	10	22	06:37	45.80	26.40	125.	6.5
1940	11	10	01:39	45.80	26.70	150.	7.7
1945	3	12	20:51	45.60	26.40	125.	6.1
1945	9	7	15:48	45.90	26.50	80.	6.8
1945	12	9	06:08	45.70	26.80	80.	6.5
1946	11	3	18:47	45.60	26.30	140.	6.0
1948	5	29	04:48	45.80	26.50	130.	6.3
1973	8	20	15:18	45.70	26.40	73.	6.0
1976	10	1	17:50	45.60	26.40	146.	6.0
1977	3	4	19:21	45.70	26.70	94.	7.4
1986	8	30	21:28	45.50	26.40	131.	7.1
1990	5	30	10:40	45.80	26.80	90.	6.9
1990	5	31	00:17	45.80	26.90	86.	6.4

Carpathians arc, are given in Figure 3. They are close to the similar distribution presented by TRIFU (1990). The seismic moment rate ($\dot{M}_0 = 1.2 \times 10^{19}$ Nm/year) is close to the estimation of ISMAIL-ZADEH *et al.* (1996), who demonstrate that such a high value could not be explained by a pure phase-transition in the Vrancea slab.

The focal mechanism of the Vrancea intermediate-depth earthquakes has been the subject of many studies. The reverse faulting with principal T axis nearly vertical and P axis nearly horizontal characterizes all the major events ($M_w > 6$) and more than 90% of the studied events, regardless of their magnitude (ENESCU, 1980; ENESCU and ZUGRAVESCU, 1990; ONCESCU and TRIFU, 1987). For the fault plane orientation, two typical solutions are obtained: (I) the fault plane oriented mainly in a NE–SW direction and P axis perpendicular to the mountain arc; and (II) the fault plane oriented mainly in a NW–SE direction and the P axis parallel to the mountain arc. A spectacular jump from one solution to the other was noticed in May 1990 when, within a 12-hour interval, a first shock ($M_w = 6.9$) with mechanism (I) was followed by a second shock ($M_w = 6.3$) with mechanism (II). The available

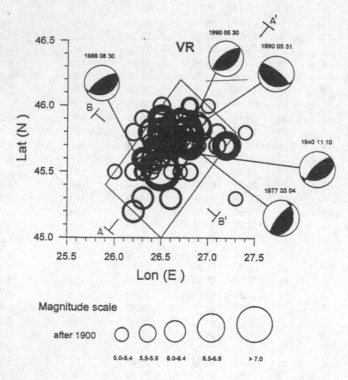

Figure 2

Epicenter distribution of Vrancea intermediate-depth earthquakes ($M_w \geq 5$) which occurred after 1900. The fault plane solutions of the five major events of November 10, 1940 ($M_w = 7.7$), March 4, ($M_w = 7.4$), August 30, 1986 ($M_w = 7.1$), May 30, 1990 ($M_w = 6.9$) and May 31, 1990 ($M_w = 6.3$) are plotted.

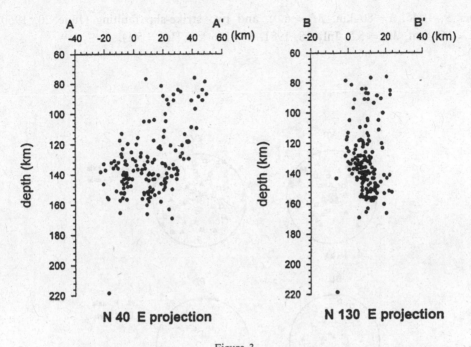

Figure 3

Cross sections of the foci depth distribution in two vertical planes: parallel (N40°E) and perpendicular (N130°E) oriented relative to the Carpathians arc. Intermediate-depth earthquakes with $M_w \geq 4$, recorded between January 1980 and August 1998 (162 events), are considered.

macroseismic and instrumental data indicate for all events with $M_w > 7$ the type (I) solution.

The stereographic projection on the lower hemisphere of the distribution of P and T principal axes is given in Figure 4. For the Vrancea subcrustal earthquakes only magnitudes $M_w \geq 5$ are considered. A clear predominant horizontal compressive stress regime is outlined in agreement with all previous studies of Vrancea earthquakes (e.g., ONCESCU and TRIFU, 1987; ENESCU and ZUGRAVESCU, 1990). The plane striking in the NE–SW direction and dipping towards NW (type I) is considered a rupture plane, in agreement with the spatial distribution of the aftershocks, teleseismic waveform inversion and source directivity analysis (MÜLLER et al., 1978; TRIFU and ONCESCU, 1987; TAVERA, 1991). It is the most commonly observed over the entire magnitude range. The type II case, with the rupture plane oriented almost perpendicular to the rupture plane of type I events, is observed for a smaller number of earthquakes and not for the largest shocks ($M_w > 7$). There are a few solutions of normal or strike-slip faulting that occur at the upper and lower edges of the subducted volume: among the 28 fault plane solutions of earthquakes with $M_w \geq 5$, only one represents normal faulting (December 9, 1945, $h = 80$ km, $M_w = 6.0$), and two strike-slip faulting (June 20, 1950, $h = 160$ km, $M_w = 5.5$; July 18, 1981, $h = 166$ km, $M_w = 5.1$).

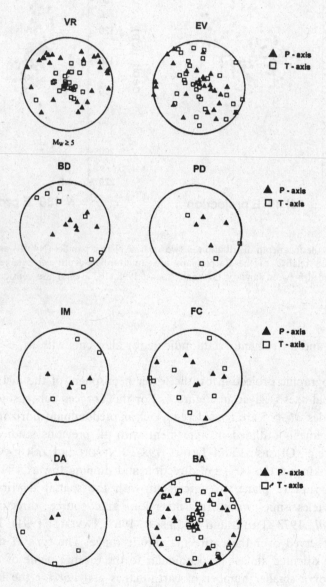

Figure 4
Equal area lower hemisphere projection of the P and T axes of the fault plane solutions in different seismogenic zones. For the Vrancea intermediate-depth earthquakes only events with $M_w \geq 5$ are considered.

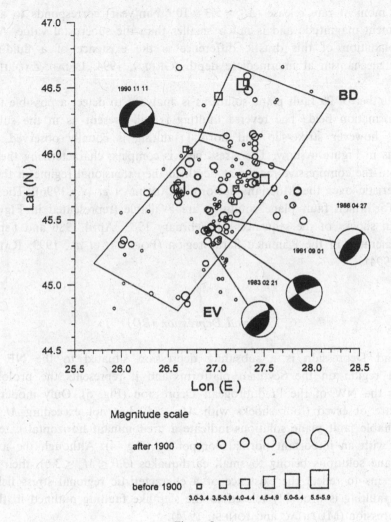

Figure 5
Epicenter distribution of the earthquakes in the Vrancea crustal zone (EV) and Bârlad Depression (BD).
Representative fault plane solutions for the four largest instrumentally recorded earthquakes are plotted.

East Vrancea Zone (EV)

As can be seen in Figure 5, the shallow seismicity associated with the Vrancea subducting process spreads more diffusely eastward relative to the Carpathians arc bend, in the strip delimited by the Peceneaga-Camena fault to the north and Intramoesian fault to the south (so-called Black Sea subplate). It consists of only moderate-size earthquakes, not exceeding magnitude 5.6. Bursts of seismic activity are relatively common in the east (Râmnicu Sarat region) and north (Vrâncioaia region).

The moment rate release ($\dot{M}_0 = 5.3 \times 10^{15}$ Nm/year) corresponds to a $M_w = 4.4$ moment magnitude and is much smaller than the subcrustal value. A possible explanation of this drastic difference is the existence of a fluid-assisted faulting mechanism at intermediate depth (GREEN, 1994; ISMAIL-ZADEH *et al.*, 1996).

A number of 29 fault plane solutions is analyzed to detect a possible trend in the deformation field. The reverse faulting is still present as in the subcrustal domain, however strike-slip and normal faulting is equally observed, as the diagrams in Figure 4 show. The stress field is complex, characterizing the transition from the compressive regime at depth to the extensional regime in the crust, characteristic over the Moesian platform (RADULIAN *et al.*, 1996). The largest events for which fault plane solutions are available (represented in Fig. 5) are the main shocks of the sequences of February 1983, April 1986 and September 1991 generated in the Râmnicu Sărat region (POPESCU *et al.*, 1993; RADULIAN *et al.*, 1994).

Bârlad Depression (BD)

Bârlad Depression is a subsiding depression situated to the NE of the Vrancea region on the Scythian platform, and it represents the prolongation towards the NW of the Predobrogean Depression (Fig. 5). Only moderate-size events are observed (four shocks with $M_w > 5.0$, but not exceeding $M_w = 5.6$). All available fault plane solutions indicate a predominant horizontal extensional regime, with an important normal component (Fig. 4). Although the available fault plane solutions belong to small earthquakes ($3.0 \leq M_w \leq 3.6$), their consistency seems to reflect the existence of a characteristic regional stress field. The normal faulting is probably related to the step-like faulting outlined in the Bârlad Depression (MUTIHAC and IONESI, 1974).

Predobrogean Depression (PD)

This seismogenic zone belongs to the southern margin of the Predobrogean Depression marked by the Sfântul Gheorghe fault. Roughly, the seismicity (Fig. 6) and focal mechanism (Fig. 4) characteristics are similar to those mentioned for the Bârlad Depression: the moderate seismic activity ($M_w \leq 5.3$), clustered especially along the Sfântul Gheorghe fault, and the extensional regime of the deformation field. In our opinion, this consistently reflects the affiliation of the two zones to the same tectonic unit (Scythian platform). From this point of view, they could be alternatively merged in a single seismogenic zone.

Intramoesian Fault (IM)

The Intramoesian fault crosses the Moesian platform in a SE–NW direction, separating two distinct sectors with different constitution and structure of the basement. Although it is a well-defined deep fault, reaching the base of the lithosphere (ENESCU, 1992), and extending southeast to the Anatolian fault region (SĂNDULESCU, 1984), the associated seismic activity, represented in Figure 6, is

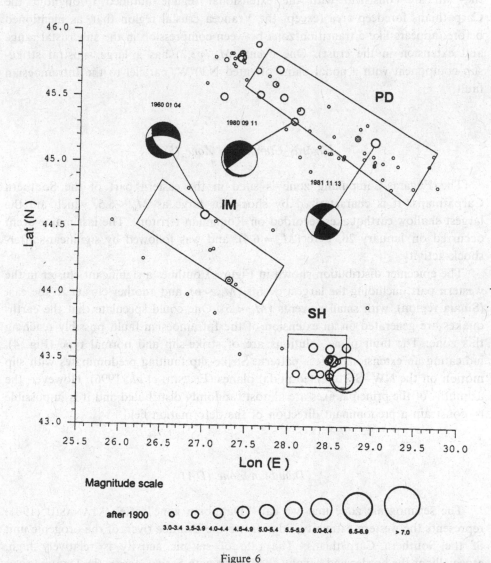

Figure 6
Epicenter distribution of the earthquakes in the Predobrogean Depression (PD), Intramoesian fault (IM) and Shabla zone (SH). Representative fault plane solutions for the three largest instrumentally recorded earthquakes are plotted.

scarce and weak, with only two events above magnitude 5. A significant increase of seismicity is observed in the Shabla region, Bulgaria (SH), where an earthquake with magnitude $M_w = 7.2$ occurred in 1901. The focal depth, whenever it can be constrained, has relatively large values ($h \sim 35$ km), suggesting active processes in the lower crust or in the upper lithosphere.

It is difficult to reach a conclusion regarding the characteristic stress pattern on the basis of the four fault plane solutions available for this area (Fig. 4), even if they all are consistent with the extensional regime outlined throughout the Carpathians foredeep area (except the Vrancea crustal region that, as mentioned before, appears like a transition zone between compression in the subcrustal range and extension in the crust). One event ($M_w = 3.2$) has a large sinistral strike-slip component with a nodal plane oriented N30°W, parallel to the Intramoesian fault.

Făgăraş-Câmpulung Zone (FC)

The Făgăraş-Câmpulung zone is sited in the eastern part of the Southern Carpathians. It is characterized by shocks as large as $M_w \sim 6.5$, which are the largest shallow earthquakes recorded on Romanian territory. The last major event occurred on January 26, 1916 ($M_w = 6.4$), and was followed by significant after-shock activity.

The epicenter distribution shown in Figure 7 outlines a significant cluster in the western part, including the largest events ($M_w \sim 6$), and another cluster to the east (Sinaia region), with smaller events ($M_w < 5$). One could speculate that the earth-quakes are generated on an extension of the Intramoesian fault, possibly reaching this zone. The fault plane solutions are of strike-slip and normal type (Fig. 4), indicating an extensional stress pattern. Strike-slip faulting predominates with slip motion on the NW–SE oriented nodal planes (ENESCU *et al.*, 1996). However, the azimuths of the principal axes are almost randomly distributed and it is impossible to constrain a predominant direction of the deformation field.

Danubian Zone (DA)

The seismogenic zone that we call "Danubian zone" after ATANASIU (1961) represents the western extremity, adjacent to the Danube river, of the orogenic unit of the Southern Carpathians. The rate of seismic activity is relatively high, especially at the border and beyond the border with Serbia, across the Danube river (Fig. 8). The magnitude does not exceed 5.6.

The fault plane solutions are available for three earthquakes (the fault plane

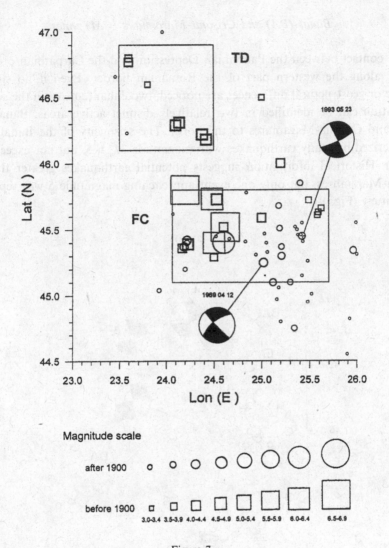

Figure 7
Epicenter distribution of the earthquakes in the Făgăraş-Câmpulung (FC) and Transylvanian Depression (TD). Representative fault plane solutions for the two largest instrumentally recorded earthquakes are plotted.

solution for the largest, M_w 5.6 event, of July 18, 1991, is plotted on Fig. 8) and indicate normal faulting with the T axis striking roughly N–S (Fig. 4), in agreement with the general extensional stress regime in the Southern Carpathians (ONCESCU et al., 1988; RADULIAN et al., 1996). However, for a firm conclusion, more fault plane solutions are required.

Banat (BA) and Crişana-Maramureş (CM) zones

The contact between the Panonnian Depression and the Carpathian orogen lies entirely along the western part of the Romanian border. Even if no significant tectonic or geostructural differences are noticed, two enhancements in the seismicity distribution can be identified in two relatively distinct active areas: Banat to the south, and Crişana-Maramureş to the north. The seismicity of the Banat zone is characterized by many earthquakes with magnitude $M_w > 5$, but not exceeding 5.6 (Fig. 8). Historical information suggests potential earthquakes greater than 6 in Crişana-Maramureş, but only one event approaching magnitude 5 was reported in this century (Fig. 9).

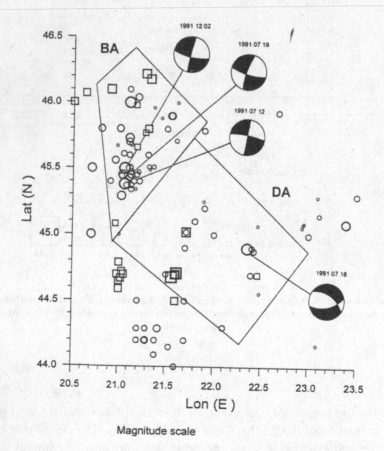

Figure 8
Epicenter distribution of the earthquakes in the Danubian (DA) and Banat (BA) zones. Representative fault plane solutions for the four largest instrumentally recorded earthquakes are plotted.

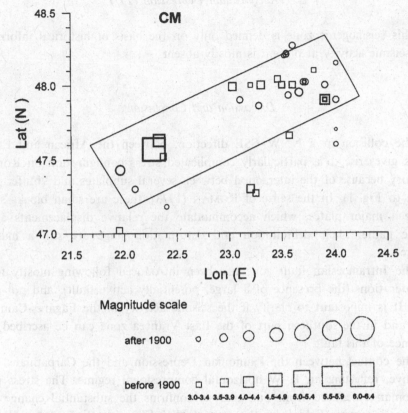

Figure 9
Epicenter distribution of the earthquakes in the Crişana-Maramureş (CM) zone.

In contrast with what we obtained in the foredeep area of the Carpathians (except Vrancea zone) and in the Southern Carpathians, where, despite randomness in the fault plane solutions, not a single reverse faulting is noticed, here reverse and strike-slip faulting predominates. The orientation of the P axes is not well constrained, however a regional horizontal compression field with an E–W direction is quite consistently outlined by the principal axes diagram (Fig. 4). As shown by RADULIAN et al. (1996), the fault plane solutions available along the Eastern Pannonian margin and in the Eastern Carpathians area (although for scarce and weak events) suggest a compressive character of the stress field. This is in agreement with the work of GRÜNTHAL and STROMEYER (1992), who pointed out that the approximately radial pattern of the extensional regime in the Pannonian Basin implies E–W compression east of the basin, in the intra-Carpathian region.

Transylvanian Depression (TD)

This seismogenic zone is defined only on the basis of historical information. The seismic activity at present is mostly absent.

Discussion and Conclusions

The collision in a NNW–SSE direction, between the African and Eurasian plates gives rise to a particularly complicated stress configuration on Romanian territory because of the interaction between several subplates and "buffer plates" (inset to Fig. 1). In the sense of ROMAN (1973), these are small blocks situated between major plates, which accommodate the relative displacements as in a puzzle game. Their delimitation and deformation stage are still a matter of debate.

The Intramoesian fault zone has been introduced following mostly tectonic considerations (the presence of a large, potentially active fault), and not seismic data. It is important to clarify if the seismic activity in the Făgăraş-Câmpulung zone and in the southern part of the East Vrancea zone can be ascribed to the presence of this fault.

The contact between the Pannonian Depression and the Carpathians orogen is active, reflecting an E–W horizontal compresional regime. The stress pattern in Romania deduced from our analysis confirms the substantial change of the compression stress field from the general NW–SE orientation (due to the push of the African plate) to the E–W orientation, as observed by GRÜNTHAL and STROMEYER (1992) on a reduced data set (including not a single fault plane solution from the Banat region, for example). The counterclockwise rotation of the Italian peninsula to the NE and the extension in the Pannonian Depression are additional plate tectonic motions which probably contribute to this particular dominant E–W oriented stress pattern.

The horizontal compression is no longer present in the outer part of the Carpathians, to the southeast (in Moesian and Scythian platforms). The compression field seems to be "consumed" at subcrustal depth by a very confined fluid-assisted faulting process (GREEN and HOUSTON, 1995; ISMAIL-ZADEH et al., 1996). If continent–continent collisional forces continue to act in the Vrancea area (CONSTANTINESCU and ENESCU, 1984; FUCHS et al., 1979), it is difficult to correlate them with the available seismicity and focal mechanism data in the crust. The seismic activity eastward of the Carpathians arc bend, along the Intramoesian and Peceneaga-Camena major faults, is too weak to justify an active northwestward movement of the Black Sea subplate, as assumed by several authors (e.g., AIRINEI, 1977; CONSTANTINESCU and ENESCU, 1984; ENESCU and ENESCU, 1993). The clustered shallow seismicity adjacent to the Vrancea region rather bears relation with the flexure process of the subducting plate beneath Vrancea. The mechanism of the dehydration, possible in the intermediate-depth range (GREEN, 1994; ISMAIL-ZADEH et al., 1996), could explain the contrast between the high subcrustal deformation rate and considerably lower value in the crust.

Although the number of fault plane solutions is not always statistically significant, and the size of the instrumentally recorded earthquakes is generally moderate, the consistent trends for some seismogenic zones, outlined by our analysis, are considered as valid preliminary indications of possible regional characteristics to be used in seismic hazard evaluations. However, the present study is restrained to a simple interpretation of the earthquake mechanisms, and a formalized relationship between tectonics and seismicity is critical in future work to infer the principal stress axes from the principal strain axes.

Acknowledgements

This research has been made possible by the NATO Linkage Grant ENVIR.LG 960916, by MURST (40% and 60%), by COPERNICUS Project, ERBCIPACT 94-0238, and is a contribution to UNESCO-IGCP Project 414 "Realistic Modelling of Seismic Input for Megacities and Large Urban Areas."

REFERENCES

AIRINEI, S. (1977), Lithospheric Microplates on the Romanian Territory Reflected by Regional Gravity Anomalies (in Romanian), St. Cerc. Geol., Geogr., Geofiz. 15, 19–30.

ATANASIU, I., Earthquakes of Romania (in Romanian) (Academy Publishing House, Bucharest 1961).

CONSTANTINESCU, L., and ENESCU, D. (1984), A Tentative Approach to Possibly Explaining the Occurrence of the Vrancea Earthquakes, Rev. Roum. Géol., Géophys., Géogr., Ser Géophys. 28, 19–32.

CONSTANTINESCU, L. and MÁRZA, V. (1980), A Computer-compiled and Computer-oriented Catalogue of Romania's Earthquakes During a Millennium (AD 984–1979), Rev. Roum. Géol., Géophys., Géogr., Ser Géophys. 24, 171–191, Bucharest.

CONSTANTINESCU, L., CONSTANTINESCU, P., CORNEA, I., and LĂZĂRESCU, V. (1976), Recent Seismic Information on the Lithosphere in Romania, Rev. Roum. Géol., Géophys., Géogr., Ser Géophys. 20, 33–40.

ENESCU, D. (1980), Contributions to the Knowledge of the Focal Mechanism of the Vrancea Strong Earthquake of March 4, 1977, Rev. Roum. Géol., Géophys., Géogr., Ser Géophys. 24, 3–18.

ENESCU, D. (1992), Lithosphere Structure in Romania. I. Lithosphere Thickness and Average Velocities of Seismic Waves P and S. Compression with Other Geophysical Data, Rev. Roum. Phys. 37, 623–639.

ENESCU, D., and ZUGRAVESCU, D. (1990), *Geodynamic Considerations Regarding the Eastern Carpathians Arc Bend, Based on Studies on Vrancea Earthquakes*, Rev. Roum. Géophysique *34*, 17–34.

ENESCU, D., and ENESCU, B. D. (1993), *Contributions to the Knowledge of the Genesis of the Vrancea (Romania) Earthquakes*, Romanian Reports in Physics *45*, 777–796.

ENESCU, D., POPESCU, E., and RADULIAN, M. (1996), *Source Characteristics of the Sinaia (Romania) Sequence of May–June 1993*, Tectonophysics *261*, 39–49.

FUCHS, K., BONJER, K. P., BOCK, G., CORNEA, I., RADU, C., ENESCU, D., JIANU, D., NOURESCU, A., MERKLER, G., MOLDOVEANU, T., and TUDORACHE, G. (1979), *The Romanian Earthquake of March 4, 1977*, Tectonophysics *53*, 225–247.

GREEN, H. W. II (1994), *Solving the Paradox of Deep Earthquakes*, Scientific American, 50–57.

GREEN, H. W. II, and HOUSTON, H. (1995), *The Mechanics of Deep Earthquakes*, Ann. Rev. Earth Planet. Sci. *23*, 169–213.

GRÜNTHAL, G., and STROMEYER, D. (1992), *The Recent Crustal Stress Field in Central Europe: Trajectories and Finite Element Modeling*, J. Geophys. Res. *97*, 11,805–11,820.

ISMAIL-ZADEH, A. T. PANZA, G. F., and NAIMARK, B. M. (1996), *Stress in the Descending Relic Slab Beneath Vrancea, Romania*, ICTP Internet Report IC/96/93.

KANAMORI, H. (1977), *The Energy Release in Great Earthquakes*, J. Geophys. Res. *82*, 2981–2987.

MÂNDRESCU, N., POPESCU, E., RADULIAN, M., UTALE, A., and PANZA, G. F. (1997), *Seismicity and Stress Field Characteristics for the Seismogenic Zones of Romania*, EEC Technical Report, Project CIPA CT94-0238.

MOLNAR, P. (1979), *Earthquake Recurrence Intervals and Plate Tectonics*, Bull. Seismol. Soc. Am. *69*, 115–133.

MÜLLER, G., BONJER, K.-P., STÖCKL, H., and ENESCU, D. (1978), *The Romanian Earthquake of March 4, 1977*, J. Geophys. *44*, 203–208.

MUSSON, R. M. W., *An earthquake catalogue for the Circum-Pannonian Basin*. In *Seismicity of the Carpatho-Balcan Region*, Proc. of XVth Congress of the Carpatho-Balcan Geol. Ass. (eds. Papanikolaou, D., and Papoulia, J.) (Athens 1996) pp. 233–238.

MUTIHAC, V., and IONESI, L., *Geology of Romania* (in Romanian) (Technical Press, Bucharest 1974).

ONCESCU, M. C., and TRIFU, C.-I. (1987), *Depth Variation of the Moment Tensor Principal Axes in Vrancea (Romania) Seismic Region*, Ann. Geophysicae *5B*, 149–154.

ONCESCU, M. C., ARDELEANU, L., POPESCU, E., (1988), *The State of Stress under the Meridional Carpathians*, Proc. of XXIst Gen. Ass. of ESC, Sofia, 1988, 149–154.

ONCESCU, M. C., MÂRZA, V. I., RIZESCU, M., and POPA, M. *The Romanian earthquake catalogue between 984–1996*. In *Vrancea Earthquakes: Tectonics, Hazard and Risk Mitigation* (eds. Wenzel, F., Lungu, D., and Novak, O.) (Kluwer Academic Publishers 1999), pp. 43–49.

PANZA, G. F., SOLOVIEV, A. A., and VOROBIEVA, I. A. (1997), *Numerical Modelling of Block Structure Dynamics: Application to the Vrancea Region*, Pure appl. geophys. *149*, 313–336.

POPESCU, E., BAZACLIU, O., and RADULIAN, M. (1993) *The Earthquake Sequence of Râmnicu Sarat (Romania) 31 August–1 September 1991*, Proc. of XXIIIrd Gen. Ass. of ESC, Prague, 1992, 86–89.

RADU, C., APOPEI I., and UTALE, A., *Contributions to the study of the seismicity of Romania* (in Romanian). In *Progrese in Fizica* Symposium (Cluj-Napoca 1980).

RADULIAN, M., POPESCU, E., and BAZACLIU, O. (1994), *A Statistical Analysis of the Heterogeneity of the Generation of the Earthquake Sequences in the Vrancea Crust*, Rom. J. Phys. *39*, 343–351.

RADULIAN, M., MÂNDRESCU, N., POPESCU, E., UTALE, A., and PANZA, G. F. (1996), *Seismic Activity, Stress Field and Seismogenic Zones in Romania*, ICTP Preprint IC/96/256.

ROMAN, C. (1973), *Rigid Plates, Buffer Plates and Sub-plates, Comment on 'Active Tectonics of the Mediterranean Region' by D. P. McKenzie*, Geophys. J. R. Astron. Soc. *33*, 369–373.

SĂNDULESCU, M., *Geotectonics of Romania* (in Romanian) (Technical Press, Bucharest 1984).

TAVERA, J., *Etude des mécanisemes focaux de gros séismes et sismicité dans la région de Vrancea-Roumanie*, Rapport de stage de recherche, Institut de Physique du Globe, Paris, 1991.

TRIFU, C.-I., and ONCESCU, M. (1987), *Fault Geometry of August 30, 1986 Vrancea Earthquake*, Annales Geophysicae *5B*, 727–730.

TRIFU, C.-I. (1990), *Detailed Configuration of Intermediate Seismicity in the Vrancea Region*, Rev. de Geofisica *46*, 33–40.

(Received April 13, 1998, revised September 22, 1998, accepted May 31, 1999)

 To access this journal online:
http://www.birkhauser.ch

Pure appl. geophys. 157 (2000) 79–95
0033–4553/00/020079–17 $ 1.50 + 0.20/0

Pure and Applied Geophysics

Identification of Future Earthquake Sources in the Carpatho-Balkan Orogenic Belt Using Morphostructural Criteria

A. I. GORSHKOV,[1] I. V. KUZNETSOV,[1] G. F. PANZA[2,3] and A. A. SOLOVIEV[1]

Abstract—The Carpatho-Balkan mountain belt, the most seismic zone of the Circum-Pannonian region, is studied with a goal to identify sites where shallow earthquakes with $M \geq 6.5$ may occur. The study is based on the assumption that strong earthquakes associate with disjunctive nodes that are formed around the junctions of lineaments. In the study region, lineaments and disjunctive nodes were defined by a morphostructural zonation method. The morphostructural map compiled at the scale of 1:1,000,000 shows a hierarchical system of homogeneous blocks, the network of morphostructural lineaments and the loci of disjunctive nodes. Shallow earthquakes with $M_s \geq 5.0$ recorded in the region were found to be nucleated at the mapped nodes. In the Carpatho-Balkan mountain belt, the nodes where earthquakes with $M \geq 6.5$ may occur have been identified using morphostructural criteria of high seismicity, previously derived from pattern-recognition of potential seismic nodes in the Pamirs-Tien Shan region. In total, 64 of the 165 nodes mapped in the studied region have been defined to be prone to earthquakes of $M \geq 6.5$. These 64 nodes include the seven where earthquakes of $M_s \geq 6.0$ already took place.

Key words: Block-structure, morphostructural zonation, criteria of seismicity, Carpatho-Balkan mountain belt.

Introduction

The Circum-Pannonian region is a well-defined seismo-active region in Europe where a single strong earthquake may trigger a major catastrophe due to high density of population and the presence of high-risk objects such as nuclear power plants and chemical plants. In the recent past and currently, seismic hazard in this region is intensively studies in the framework of international multidisciplinary research projects (e.g., QUANTITATIVE SEISMIC ZONING OF THE CIRCUM-PANNONIAN REGION, 1996, 1997). In the Circum-Pannonian region, most strong earthquakes occur in the mountain belts surrounding the young sedimentary basins located centrally in the region.

[1] International Institute of Earthquake Prediction Theory and Mathematical Geophysics, Russian Academy of Sciences, Moscow, Russia.
[2] Department of Earth' Sciences, University of Trieste, Italy.
[3] The Abdus Salam International Center for Theoretical Physics, Trieste, Italy.

In this work, we study the Carpathians and its structural extension, the Balkanides, with a goal to identify sites where strong earthquakes may occur. Both mountain systems are considered as a single large-scale geotectonic unit within the framework of the morphostructural zonation method.

The study is based on the assumption that strong earthquakes associate with disjunctive nodes, specific structures that are formed around the junctions of mobile zones called lineaments. The fact that earthquakes are nucleated in the nodes was first established from observations in the Pamirs and Tien Shan (GEL'FAND et al., 1972). The nodes were mapped by a morphostructural zonation method that is thoroughly described by ALEXEEVSKAYA et al. (1977). Pattern recognition was used in order to identify nodes where strong earthquakes may occur (GEL'FAND et al., 1976). The physical mechanism of nodes formation was first introduced by MCKENZIE and MORGAN (1969). The problem why strong earthquakes are nucleated around fault junctions was discussed by GABRIELOV et al. (1996).

During the two last decades many regions of the world have been studied using morphostructural zonation in combination with pattern recognition for identification of earthquake-prone areas (CAPUTO et al., 1980; CISTERNAS et al., 1985; GEL'FAND et al., 1972, 1976; GVISHIANI et al., 1986, 1987, 1988; BHATIA et al., 1992; GORSHKOV et al., 1991, 1994). These studies demonstrated that strong and moderate earthquakes are nucleated in the nodes outlined by morphostructural zonation. The non-randomness of earthquake nucleation in the nodes was proved statistically by an especially designed method (GVISHIANI and SOLOVIEV, 1980). Earthquakes are not nucleated in all the nodes but in some of them fitting certain criteria defined by pattern recognition. These criteria of seismicity are partly similar for different regions (see the above references). Therefore, criteria characteristic to the seismic nodes in one region can be used for identification of potential seismic nodes in an other region with a relatively similar geodynamical environment.

In this work, we attempted to identify high-seismic nodes ($M \geq 6.5$) in the Carpatho-Balkan mountain belt applying the criteria of high seismicity defined by pattern recognition in the Pamirs and Tien Shan (KOSSOBOKOV, 1983). In a given case, pattern recognition is inapplicable for this purpose because the number of strong earthquakes recorded in the Carpatho-Balkan mountain belt is absolutely insufficient for the learning stage of pattern recognition.

Methodology

The mapping of nodes is based on the methodology called morphostructural zonation (MSZ). MSZ is based on ideas widely used in the morphostructural analysis of block-structure of the crust (RANTSMAN, 1979). MSZ means the division of a territory into a system of hierarchically-ordered areas characterised by a definite degree of uniformity of the morphostructures. By a MSZ three types of morphostructures are distinguished: (1) blocks of different ranks; (2) their boundaries represented by the linear zones (lineaments); (3) sites where linear zones intersect, called disjunctive nodes.

The following characteristics of topography are the subject of the analysis: (1) elevation and orientation of large topographic forms, and its variations; (2) drainage pattern and its variations; (3) linear elements of topography such as rivers, ravines, escarpments, bottom edges of the slopes of terraces and valleys.

At the first stage of the analysis homogeneous blocks are delineated. An area is defined as block if: (1) within this area elevations and orientations of large topographic forms are homogeneous; (2) drainage pattern is uniform for the entire area (e.g., big rivers are nearly parallel to each other).

At the second stage, zones of lineaments are identified at sites where: (1) elevation of topography is sharply changed (1/10 of average elevation within an area); (2) there is a narrow strip of straight segments of linear elements of topography which has a uniform strike; (3) there are sharp turns of predominant strike of large topographic forms (by 30°).

MSZ is hierarchically ordered and territorial units (blocks) and lineaments are assigned with ranks. Mountain countries are considered as the highest (first) rank units. They are divided into second rank units called megablocks. Megablocks are further subdivided into units of third rank called blocks. Neighbouring blocks should differ at least in one of the above characteristics. Megablocks are territories within which all the characteristics of topography are similar or change with a common regularity.

The rank of lineament is also defined as first, second and third, depending on the rank of the territorial unit limited by the lineament. Actually, lineaments are zones of tectonic deformations. Depending on their orientation with respect to the general stretch of mountain belt, two types of boundaries are distinguished—longitudinal and transverse lineaments. Longitudinal ones are generally parallel to the predominant strike of large topographic elements. They occur along the boundaries of these elements, separating the relatively uplifted areas from those of lower elevation. Longitudinal lineaments are also characterised by contrasting types of topography on both sides of the lineament and include long segments of the prominent faults. Transverse lineaments are oriented across or at an angle to the predominant stretch of the mountain chain. Normally, they appear on the earth's surface discontinuously and are represented by tectonic escarpments, rectilinear segments of river valleys and fault segments.

The nodes are formed at sites where block boundaries of different orientation intersect, and are characterized by particularly intensive fracturing and contrasting neotectonic movements with a mosaic pattern of structure and topography resulting (RANTSMAN, 1979; GVISHIANI et al., 1998).

It should be emphasised that morphostructural zonation is performed with no connection with seismicity data in a studied region.

Morphostructural Zonation of the Carpatho-Balkan Mountain Belt

The morphostructural map of the study region, represented in Figure 1, is based on the joint analysis of topographic, tectonic, geological maps and satellite photos.

The Carpatho-Balkan mountain system consists of the two mountain chains, the Carpathians and the Balkanides. In MSZ each of them is considered as the first rank territorial unit.

The Carpathians (marked by "*C*" in Fig. 1). This is one of the most complex segments of the Europe Alpine system, constituting an extremely complicated structure. The recent geodynamics in this area is controlled by the ongoing

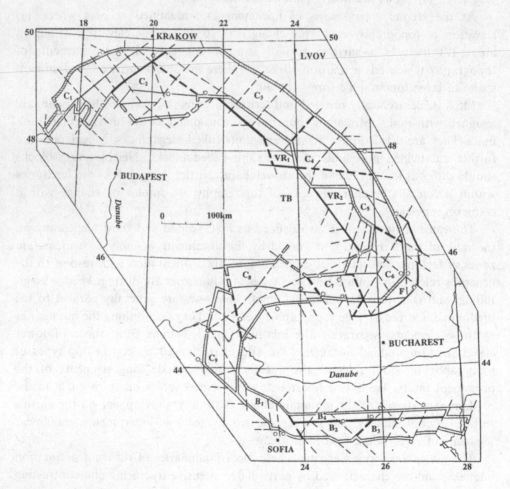

Figure 1

Morphostructural zonation map of the Carpatho-Balkan mountain belt. Lines are lineaments; double lines—first rank lineaments; bold lines—second rank lineaments, thin lines—third rank lineaments; continuous lines—longitudinal lineaments; discontinuous lines—transverse lineaments. Circles are epicentres of earthquakes with $M \geq 5.0$.

interaction of numerous lithosphere plates and subplates (RADULESCU and SANDULESCU, 1973), and the mountain system represents a continental collision zone, developed in a compressional regime (REBAI et al., 1992; RADULIAN et al., 1996b). On the earth's surface the mountain domain is represented by an extremely extended arc-like ridge, the elevation of which does not exceed 2,500 m.

The Carpathians are bordered by first rank lineaments corresponding to the prominent faults. To the east and to the south, the mountain domain is separated by first rank longitudinal lineament from the Carpathian foredeep, which is filled with Neogen-Quaternary molasses. Most of the folded and uplifted part of the foredeep is included in the map as the megablock which is marked by "F" in Figure 1. To the west, the Carpathians border the Transylvanian basin (marked by "TB" in Fig. 1), the Oas-Tibles-Gutai and the Gurghiu-Harghta volcanic ridges (marked by "VR_1" and "VR_2" in Fig. 1). These relatively elevated ridges composed of andesitic rocks belong to the Transylvanian basin. Because of this, they were excluded from the Carpathians and assigned to two isolated megablocks, VR_1 and VR_2. Due to the same reasons the southern flank of the Western Carpathians, composed of young volcanic rocks, is excluded from the mountain country.

To the south, transverse first rank lineament traced along the Timok fault (KHAIN, 1978) separates the Carpathians from the Balkanides. To the west, a system of first-rank lineaments delimits the Carpathians from the Vienna basin and the Pannonian basin.

As a result of the morphostructural zonation, the mountain domain has been divided into nine megablocks (marked by C_1–C_9 in Fig. 1). They differ in (1) elevation, (2) predominant orientation of ranges composing the mountain chain and (3) combination of large topographic forms.

Megablock C_1 which includes most of the western part of the Carpathians, comprises a system of nearly parallel ridges with NE-SW orientation. The elevation of these ridges increases from north to south. These ridges are separated from each other by a longitudinal lineament of third rank. The southeastern boundary of the megablock is a longitudinal lineament of second rank passing along the tectonic boundary between the Outer and Inner Carpathians. Transverse lineaments with NW-SE orientation are characteristic to the megablock.

Most of the uplifted part of the Western Carpathians, the Tatras, is assigned to megablock C_2. The northwestern boundary of the megablock, a transverse second rank lineament, follows the Beskidy mountain, whose elevation is considerably smaller than the Tatras. Because of the difference of elevation, the High Tatras and the Low Tatras are separated by a longitudinal third rank lineament running along the Vag River. The transverse lineaments are oriented both NW-SE and NE-SW.

Megablock C_3 occupies an intermediate position between the Western and the Eastern Carpathians. The megablock includes the most lowered segment of the Carpathians. Unlike megablocks C_1 and C_2, the mountain chain within the megablock is oriented NW-SE. The elevation of topography within the megablock does not exceed 1,000 m. The NE-SW orientation is characteristic for the transverse lineaments within the megablock. The southeastern boundary of the megablock, a transverse second rank lineament, is traced along the Teresva river. Along this boundary the elevation of the mountain chain changes sharply.

Megablock C_4 includes the widest and most uplifted part of the Eastern Carpathians and consists of two nearly parallel ridges separated by a longitudinal third rank lineament that is traced along the longitudinal river valleys. The southern boundary of the megablock, a transverse second rank lineament of nearly W-E orientation, follows the Moldova River and corresponds to the Dragos-Voda fault (PATRASCU, 1993).

Megablock C_5 is characterised by the NW-SE orientation of ridges, whose elevation gradually decreases from north to south. A longitudinal third rank lineament separates two nearly parallel ranges of different altitude. The southern boundary of the megablock, a transverse second rank lineament of NW-SE orientation, corresponds to the Trotus fault (LINZER *et al.*, 1998).

In megablock C_6, the mountain chain is oriented in the N-S direction. The structure of this Carpathians' segment is complicated by the Brashov basin, young intermountain depression. This indicates extremely contrasting neotectonic movements operating in the megablock. Transverse third rank lineaments reveal a different orientation. Their trend is in agreement with the orientation of the deep-seated fractures detected from geophysical investigations (CHEKUNOV *et al.*, 1993).

Megablocks C_7 and C_8 include the South Carpathians, the most uplifted (2,500 m) and consolidated part of the mountain chain. Both of them are characterised by similar elevation of topography and strike of the mountain chain. They are separated to account for the difference in orography. Megablock C_7 is represented by a single consolidated range, while megablock C_8 consists of many ranges of different altitude, separated by river valleys that define third rank lineaments. The boundary between these megablocks is a second rank transverse lineament passing along the Olt river. This lineament approximately corresponds to the extension of the South Transylvanian fault in the SE direction (LINZER *et al.*, 1998). In megablock C_8, a third rank longitudinal lineament of near W-E orientation corresponds to the South Carpathian fault (LINZER, 1996) and NE-SW third rank lineament is traced in agreement with the Cerna-Jiu fault (RATSCHBACHER *et al.*, 1993). In both megablocks, third rank transverse lineaments also reveal NW-SE orientation.

Megablock C_9 is a transition zone between the Carpathians and the Balkanides. The orientation of the mountain chain is nearly N-S and the elevation of topography is considerably lower with respect to the neighbouring megablocks.

Apart from the different elevation and orientation of the mountain chain within each outlined megablock, they differ in the rate of recent vertical surface movements. In accordance with the map of recent vertical movements in the Carpathians and the Pannonian basin (JOO, 1992), in megablock C_1 recent vertical movements are negative. In all other megablocks, these movements are positive although the rate of uplifting varies significantly from one megablock to the other.

The Balkanides (marked by "*B*" in Fig. 1). This mountain domain is a part of the Carpatho-Balkan arc. To the north, it borders the Moesian platform. In the MSZ map this boundary is represented by a first rank lineament corresponding to the For-Balkan fault (VELICHKOVA and SOKEROVA, 1980). To the south, a first rank lineament separates the relatively high Balkanides from the Srednegorie, characterised by lower elevation. The zone of this lineament corresponds to the Sub-Balkan fault (VELICHKOVA and SOKEROVA, 1980). The western limit of the Stara Planina range, a transverse first rank lineament, is traced in agreement with the Black Sea cryptostructure (VELICHKOVA and SOKEROVA, 1980).

The mountain domain is subdivided into four megablocks due to the differences in elevation and orientation of the Stara Planina range.

In megablock B_1, the axis of the Stara Planina range is oriented NW-SE. Transverse third rank lineaments are traced in a nearly N-S direction along river valleys of the same orientation. The western boundary of the megablock is a second rank transverse lineament traced along the Vratsa strike-slip fault and its extension in the NW-SE direction.

Megablocks B_2 and B_3 include the axial zone of the Stara Planina range. They differ from each other in the altitude of the range which is higher in megablock B_2. The boundary between them is a second transverse lineament traced NE-SW at the site where the altitude of the range sharply decreases from 2,300 m to 1,500 m.

Megablock B_4 occupies the lowered part of the northern flank of the Stara Planina range. Its southern boundary corresponds to the Stara Planina frontal line (VELICHKOVA and SOKEROVA, 1980).

Transverse second rank lineaments reveal both NW-SE and NE-SW orientations. Lineaments with NE-SW orientation are in agreement with transcurrent faults defined by geological and geophysical methods (BONCHEV, 1987; BOTEV *et al.*, 1988). A system of NW-SE lineaments in the Balkan region, including the Balkanides, has been identified on the basis of satellite data analysis (KATSKOV and STOICHEV, 1976).

In summary, the map presents a network of lineaments and a hierarchical system of blocks within the entire Carpatho-Balkan mountain chain. Such a map may be considered as a model of block-structure of the region. Relative movements and interaction of the defined blocks may cause the shallow seismicity in the region.

Nodes and Earthquakes

Totally, 165 intersections of lineaments were identified in the study region. Since the morphostructural map was compiled on the basis of cartographic sources without field research, natural boundaries of the nodes were outlined. Previously, in the Caucasus, the nodes were mapped during the field work and their dimensions had been established to be of 40–60 km by 20–40 km (GVISHIANI *et al.*, 1986, 1988). Thus the areas within the radius of 20–30 km around the points of lineament intersections can be regarded as disjunctive nodes. In this work the nodes were defined as a circle of 25 km radius surrounding the points of lineament intersections. Such node dimensions are in agreement with the size of the earthquake source for the magnitude range considered in this paper. According to RIZNICHENKO (1976), the source size of an earthquake with $M = 6.0$ is 23 km of length and about 10 km of width.

In order to evaluate the correlation between the delineated nodes and the shallow seismicity recorded in the study region, we used the earthquake catalogue for the Circum-Pannonian region (QUANTITATIVE SEISMIC ZONING OF THE CIRCUM-PANNONIAN REGION, 1996). The catalogue spans the time interval from 1010 to 1993. This catalogue was selected because it was compiled within the framework of the international project, with the participation of experts from all the countries located in the Circum-Pannonian region. The catalogue is homogeneous for moderate and strong events which are of interest in this work. The NEIC catalogue (GLOBAL HYPOCENTRES DATA BASE, 1997) was analysed for the period after 1993. According to this catalogue, in this period shallow earthquakes of the considered magnitudes did not occur within the investigated territory.

In total, we have selected 36 shallow earthquakes ($h < 60$ km) with $M_s \geq 5.0$ (Table 1), located only within the limits of the mapped territory. Some large well-known events, e.g., the 1901 Shabla earthquake of $M_w = 7.2$ in Bulgaria, were not included in Table 1 because they are beyond the limits of the mapped territory. The selected earthquakes are plotted on the morphostructural map (Fig. 1). Since the epicentre locations coincide for the earthquakes of 07.09.1932 and 06.12.1937 as well as for the events of 23.07.1913 and 14.07.1914, Figure 1 shows only 34 epicentres.

Of the 36 events considered, 28 earthquakes occurred in the Carpathians and 8 took place in the Balkanides. In the Carpathians, these earthquakes are concentrated mainly in the Western and Southern Carpathians, while in the Eastern Carpathians only a few such events are recorded along the eastern boundary of the mountain chain.

As can be seen in Figure 1, the epicentres of these earthquakes are located near the intersections of lineaments. We remind that the morphostructural map was compiled less any connection with seismicity data in the studied region. Therefore, we are dealing with the two independent data sets—shallow earthquakes with $M_s \geq 5.0$ and nodes resulting from MSZ.

Table 1

Earthquakes with $M_s \geq 5.0$

Year	Month	Day	Lat.	Long.	Depth	M_s
1443	6	5	48.77	18.63	36	6.4
1453	0	0	49.00	20.50	0	5.6
1473	8	29	45.60	25.00	0	5.2
1571	4	10	45.50	24.00	0	5.2
1615	1	5	48.50	18.00	0	5.0
1784	3	18	46.10	25.60	0	5.0
1785	8	22	49.70	19.00	35	5.8
1786	2	27	49.70	18.50	40	6.6
1786	12	31	49.70	20.00	40	6.6
1818	4	25	42.80	23.30	0	6.0
1826	10	16	45.70	24.50	0	5.5
1858	9	30	42.80	23.50	0	6.6
1879	10	10	44.70	21.60	0	5.0
1892	10	14	43.30	27.00	0	5.2
1893	5	20	44.30	21.20	0	5.6
1894	8	31	46.80	26.60	0	6.8
1894	12	19	45.03	21.72	0	5.0
1904	4	20	48.60	17.42	0	5.0
1906	1	9	48.60	17.55	11	5.7
1909	6	19	43.00	26.50	40	5.2
1912	6	7	45.70	26.60	25	5.5
1913	6	14	43.10	25.70	30	7.0
1913	7	23	45.70	26.80	0	5.3
1914	5	26	49.10	21.53	0	5.0
1914	6	14	43.00	25.00	33	5.0
1914	7	14	45.70	26.80	0	5.3
1916	1	26	45.40	24.60	21	6.4
1926	6	29	42.60	26.70	0	5.0
1932	9	7	45.70	26.60	0	5.4
1937	12	6	45.70	26.60	0	5.0
1963	5	4	45.10	23.40	33	5.0
1969	4	12	45.31	25.12	23	5.0
1970	7	10	47.95	25.83	33	5.0
1986	2	21	43.30	25.96	10	5.5
1990	11	11	46.03	27.13	25	5.0
1991	7	18	44.90	22.41	11	5.5

The distribution of the epicentres according to their distance to the intersections of lineaments is represented in Table 2. One can see that this distance does not exceed 20 km which is in agreement with the selected dimension of the nodes. Thus the mapped nodes can be regarded as earthquake-controlling structures in the Carpatho-Balkan mountain system. We assume that future moderate and strong earthquakes will occur within the mapped nodes.

Identification of Nodes where Earthquakes with M ≥ 6.5 May Occur

For seismic hazard evaluation it is very important to identify areas where the largest earthquakes may occur. According to the data available (Table 1), the two strongest events that affected the region registered magnitude 6.8 and 7.0. One of them occurred in the Carpathians (megablock C_5) the other one took place in the Balkanides (megablok B_4). The intriguing question is whether such strong events are possible or not in other areas of the mountain belt.

To answer the question, pattern recognition can be applied to classify the nodes defined in the region, however the number of known strong earthquakes is too small to allow a satisfactory learning phase of the pattern recognition algorithm. Because of this, for the identification of the nodes within the Carpatho-Balkan mountain belt where earthquakes with $M \geq 6.5$ may occur, we used the criteria of seismic nodes defined in the Pamirs and Tien Shan. Like the studied region, the Pamirs and Tien Shan are the young mountain systems. The Pamirs belongs to the Alpine-Himalayan mobile belt. On the whole, the recent geodynamics is relatively similar in both mountain systems. They represent active continental collision zones developed in a compressional regime (REBAI *et al.*, 1992; KHAIN, 1978). We can assume that in both regions the factors controlling seismogenesis are also relatively similar, although the seismic regime is different in each of them. Strong shallow earthquakes more frequently occur in the Pamirs and Tien Shan. However at the same time, the size of the largest events is close in both regions: maximum observed magnitudes are in the neighbourhood of 7. Besides, the objects of the study, disjunctive nodes, are defined by the same methodology in both regions. We believe that this evidence provides reason enough to apply criteria of high seismicity defined in the Pamirs and Tien Shan for the identification of potential seismic nodes in the Carpatho-Balkan mountain belt.

The tested criteria have been obtained by pattern recognition using Hamming's method (KOSSOBOKOV, 1983) on the basis of the morphostructural map of the Pamirs Tien Shan region compiled by RANTSMAN (1979). In the Pamirs and Tien Shan, the nodes recognised to be prone to earthquakes with $M \geq 6.5$ possess the following characteristics:

(1) the relief energy (difference between maximum and minimum topographic altitude inside a node) must be greater than 2500 m;

(2) a node must be located inside the mountain domain;

Table 2

Distribution of the epicentres according to their distance to the intersections of lineaments

Distance to intersection	0–11 km	11–20 km
Number of epicentres	23	13

(3) the highest rank of one of the lineaments forming a node must be either one or two;

(4) the number of lineaments forming a node must be two or greater.

The criteria can be interpreted as follows. Characteristics (1) and (2) indicate the higher intensity of the neotectonic movements at the nodes. Characteristics (3) and (4) indirectly indicate the increased fragmentation of the crust within the high-seismic nodes.

According to our experience, these criteria are sufficiently reliable in order to identify nodes having high seismic potential. The nodes of the Greater Caucasus have been tested by these criteria to define those where earthquakes with $M \geq 6.5$ may occur (GVISHIANI et al., 1986). The results were proved by the 1991 Racha earthquake with $M_b = 6.8$ which happened at the node, previously recognised to be prone to events with $M \geq 6.5$ (GORSHKOV, 1993).

To test the above criteria in the studied region, the values of relevant characteristics were measured from topographic maps and morphostructural maps compiled (Fig. 1) within the areas with a radius of 25 km around points of intersection of lineaments.

In recognition of nodes in the Pamirs and Tien Shan (KOSSOBOKOV, 1983), the values of the "relief energy" have been discretized in such a way that the threshold of the discretization of this characteristic divided all the nodes into two equal parts. In the studied region, we must define the threshold for this characteristic by the same means. For this purpose, we analysed one-dimensional distribution of the values of the "relief energy" measured inside the nodes of the Carpatho-Balkan mountain belt (Fig. 2). One can see that all the nodes are divided into two equal parts by the threshold of 1050 m.

In order to reduce the number of possible errors of "false alarm" type, we made the testing criteria more severe. For this purpose, the threshold for characteristic (4) has been increased: potentially seismic nodes in the Carpatho-Balkan mountain belt must be formed by at least three lineaments. Thus the following modified criteria were tested in the Carpatho-Balkan mountain belt:

(1) the "relief energy" must be greater than 1050 m;

(2) a node must be located inside the mountain domain;

(3) the highest rank of one of the lineaments forming a node must be either one or two;

(4) the number of lineaments forming a node must be three or greater.

According to those criteria, 64 nodes (about 40%) out of the 165 defined in the area, have been recognised to be prone to earthquakes with $M \geq 6.5$ (Fig. 3). These 64 nodes include the entire seven where earthquakes with $M_s \geq 6.0$ already took place. Furthermore, five nodes of the six where earthquakes with $5.5 \leq M_s \leq 5.9$ have been recorded, have also been recognised to be the potential sites for the occurrence of stronger events. Most of the nodes containing earthquakes with $5.0 \leq M_s \leq 5.4$ have not been recognised as potential sites for events with larger magnitude.

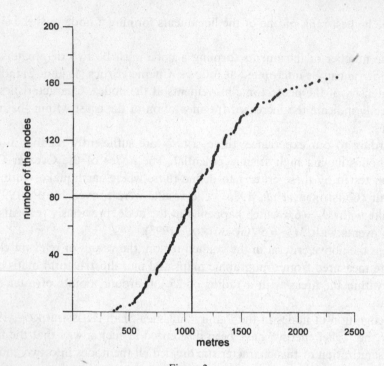

Figure 2
One-dimensional distribution of the values of the "relief energy." The values of the "relief energy" are
plotted on the horizontal axis.

Discussion and Conclusions

The recognised nodes form several clusters in both mountain systems (Fig. 3).
These clusters include practically all earthquakes of $M_s \geq 5.5$ and, in addition, they
embrace sufficiently large areas outlined zones which have high seismic potential.

In the Carpathians, high-seismic nodes are concentrated in the Western
(megablock C_2) and Southern Carpathians (megablocks C_7 and C_8), and in the
southern part of the Eastern Carpathians (megablocks C_5 and C_6). A few isolated
nodes are identified north of the Eastern Carpathians (megablock C_3). Most of the
recognised nodes are connected with a first rank lineaments and transverse lineaments of a second rank. In the Eastern and Southern Carpathians, a transverse
second rank lineaments corresponds to the large strike-slip faults: the Dragos-Voda
fault, the Trotus fault and the South Carpathian fault (see above the description of
the morphostructural map). These faults are most important in the recent kinematics of the Carpathians (LINZER *et al.*, 1998). Probably their high-level tectonic
activity defines the high-seismic potential of the recognised nodes which are
associated with these faults.

According to the seismic zoning map of Romania (STAS 11100/1–1990), e.g.,
represented in MARZA and PANTEA (1993), the maximum possible intensity in the

region is 9. The zone of this intensity embraces the Vrancea area. On our map (Figs. 1 and 3) this zone corresponds to megablock C_6 within which practically all the nodes were recognised to be the potential sites for the occurrence of earthquakes with $M \geq 6.5$. High-seismic nodes in megablocks C_7 and C_8 are correlated with zones of intensities 7 and 8. Nodes recognised in megablock C_5 fall in the zone of intensity 7.

Seismogenic zones in Romania have been recently delineated by RADULIAN *et al.* (1996a,b) on the basis of the observed shallow seismicity. Zones X, IX and Va (RADULIAN *et al.*, 1996a,b), where the recorded seismicity is most densely concentrated, include the high-seismic nodes in megablocks C_6 and C_7. High-seismic nodes

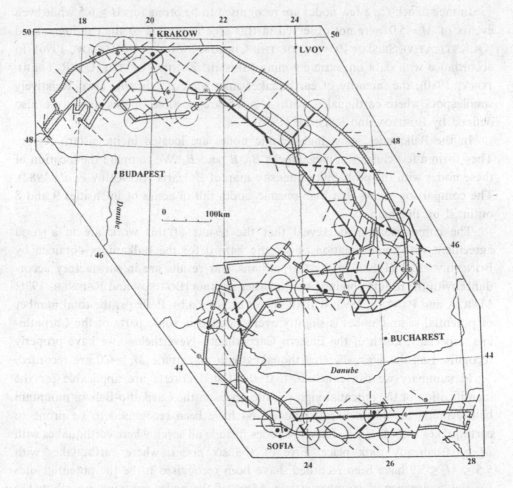

Figure 3

Nodes having potential for the occurrence of earthquakes with $M \geq 6.5$. Large circles are the nodes identified to be prone to earthquakes with $M \geq 6.5$. Small circles represent epicentres of earthquakes with $M_s \geq 6.5$ (filled circles), $M_s = [5.5-5.9]$ (crossed circles), and $M_s = [5.0-5.4]$ (open circles).

in megablocks VR_2 and C_5 correspond to the southern part of zone IV. The nodes of high seismic potential within megablock C_4 correlate with the northern part of zone IV where only low seismicity is recorded. In the comparison of these results, one should take into account that the nodes recognised in this work are the potential sites for the occurrence of strong earthquakes while the noted seismogenic zones are based only on the recorded seismicity.

In the Southern and Eastern Carpathians, sufficiently extended areas which have potential for the occurrence of earthquakes with large magnitudes (more than 6.0) have been delineated by BORISOV and REISNER (1976) on the basis of formal analysis of geological data. High-seismic nodes identified in our study are in agreement with the results of these authors.

In megablock C_3, a few nodes are recognised to be prone for $M \geq 6.5$ while even events of $M = 5.0$ were not observed in this area according to the catalogue used (QUANTITATIVE SEISMIC ZONING OF THE CIRCUM-PANNONIAN REGION, 1996). In accordance with data on seismic zoning (SEISMIC ZONING OF THE USSR TERRITORY, 1980), the intensity of earthquakes may achieve 7 in this area. Relatively small spots where earthquakes with $M \geq 6.5$ are possible in this area were also defined by BORISOV and REISNER (1976).

In the Balkanides, most high-seismic nodes are located in its eastern region. They form a few clusters in megablocks B_1, B_2 and B_4. We compare the location of these nodes with the probability intensity map of Bulgaria (BONCHEV *et al.*, 1982). The comparison shows that high-seismic nodes fall in zones of intensities 9 and 8 outlined on this map.

The comparisons made reveal that the results of this work are in a good agreement with the estimation of seismic hazard for the Balkanides obtained by BONCHEV *et al.* (1982). In the Carpathians, our results are in satisfactory accordance with the results obtained by the other authors (BORISOV and REISNER, 1976; MARZA and PANTEA, 1993; RADULIAN *et al.*, 1996a,b). Perhaps, the total number of potential seismic nodes is slightly overestimated in some parts of the Carpathians, especially, north of the Eastern Carpathians. Nevertheless, we have properly recognised all the sites where earthquakes with magnitude $M_s \geq 6.0$ are recorded.

In summary we can conclude that the tested criteria are applicable for the identification of the potential high-seismic nodes in the Carpatho-Balkan mountain belt. Among the 165 delineated nodes, 64 have been recognised to be prone to earthquakes with $M \geq 6.5$. These 64 nodes include all seven where earthquakes with $M_s \geq 6.0$ already took place. Five of the six nodes, where earthquakes with $5.5 \leq M_s \leq 5.9$ have been recorded, have been recognised to be the potential sites for the occurrence of stronger events. Most of the nodes containing earthquakes with $5.0 \leq M_s \leq 5.4$ have not been recognised as potential sites for events with larger magnitudes.

Acknowledgements

This work was partly accomplished at the International Centre for Theoretical Physics (Trieste, Italy), supported by INTAS (grant INTAS 94-0457), by the Russian Foundation for Basic Research (grant RFFI 97-05-65802), by the University of Trieste. C. Trifu and M. Radulian are gratefully acknowledged for their useful comments.

Realistic Modelling of Seismic Input for Megacities and Large Urban Areas (project 414)

REFERENCES

ALEXEEVSKAYA, M., GABRIELOV, A., GEL'FAND, I., GVISHIANI, A., and RANTSMAN, E. (1977), *Formal Morphostructural Zoning of Mountain Territories*, J. Geophys. *43*, 227–233.

CAPUTO, M., KEILIS-BOROK, V., OFICEROVA, E., RANTSMAN, E., ROTWAIN, I., and SOLOVIEV, A. (1980), *Pattern Recognition of Earthquake-prone Areas in Italy*, Phys. Earth Planet. Inter. *21*, 305–320.

CISTERNAS, A., GODEFROY, P., GVISHIANI, A., GORSHKOV, A., KOSSOBOKOV, V., LAMBERT, M., RANTSMAN, E., SALLANTIN, J., SALDANO, H., SOLOVIEV, A., and WEBER, C. (1985), *A Dual Approach to Recognition of Earthquake Prone-areas in the Western Alps*, Annales Geophysicae *3* (2), 249–270.

CHEKUNOV, A., SOLLOGUB, B., SOLLOGUB, N., KHARITONOV. O., SHLYAKHOVSKY, V., and SHCHUKIN, Yu., *Deep structure of the lithosphere in Central and South-East Europe*. In *Seismicity and Seismic Zonation of the North Eurasia. Iss. 1* (ed. Ulomov, V.) (Moscow 1993) pp. 152–161 (in Russian).

BHATIA, S. C., CHETTY, T. R. K., FILIMONOV, M., GORSHKOV, A., RANTSMAN, E., and RAO, M. N. (1992), *Identification of Potential Areas for the Occurrence of Strong Earthquakes in Himalayan Arc Region*, Proc. Indian Acad. Sci. (Earth Planet Sci.) *101* (4), 369–385.

BONCHEV, E., BUNE, V., CHRISTOSKOV, L., KOSTDINOV, V., REISNER, G., RIZHIKOVA, S., SHEBALIN, N., SHOLPO, V., and SOKEROVA, D. (1982), *A Method for Compilation of Seismic Zoning Prognostic Maps for the Territory of Bulgaria*, Geologica Balcanica *12* (2), 3–48.

BONCHEV, E. (1987), *Main Ideas in the Tectonic Synthesis of the Balkans. I. The Lithospheric Plates and the Collision Space between them*, Geologica Balcanica *17* (4), 9–20.

BORISOV, B. A., and REISNER, G. I. (1976), *Seismotectonic Prognosis of Maximum Earthquake Magnitudes in the Carpathian Region*, Izvestyia USSR Ac. Sci., Physics of the Earth. *5*, 21–31 (in Russian).

BOTEV, E., BURMAKOV, V., TREUSSA, D., and VINNIK, L. (1988), *Crust and Upper-mantle Inhomogeneities beneath the Central Part of the Balkan Region*, PEPI *51*, 198–210.

GABRIELOV, A., KEILIS-BOROK, V., and JACKSON, D. (1996), *Geometric Incompatibility in a Fault System*, Proc. Natl. Acad. Sci. USA *93*, 3838–3842.

GEL'FAND, I., GUBERMAN, Sh., IZVEKOVA, M., KEILIS-BOROK, V., and RANTSMAN, E. (1972), *Criteria of High Seismicity, Determined by Pattern Recognition*, Tectonophysics *13*, 415–422.

GEL'FAND, I., GUBERMAN, Sh., KEILIS-BOROK, V., KNOPOFF, L., PRESS, F., RANTSMAN, E., ROTWAIN, I., and SADOVSKY, A. (1976), *Pattern Recognition Applied to Earthquake Epicentres in California*, Phys. Earth Planet. Inter. *11*, 227–283.

GLOBAL HYPOCENTRES DATA BASE, CD ROM (1997), (NEIS/USGS, Denver, CO. USA).

GORSHKOV, A., ZHIDKOV, M., RANTSMAN, E., and TUMARKIN, A. (1991), *Morphostructures of the Lesser Caucaus and Places of Earthquakes, M > 5.5*, Izvestia AN USSR: Physics of the Earth. *6*, 30–38 (in Russian).

GORSHKOV, A., *Application of the results of recognition of earthquake prone-areas to seismic zonation purposes (on the example of the Caucasus)*. In *Seismicity and Seismic Zoning of Northern Eurasia* (ed. Ulomov, V.) (Moscow 1993) pp. 207–216 (in Russian).

GORSHKOV, A., TUMARKIN, A., FILIMONOV, M., and GVISHIANI, A., *Recognition of earthquake prone-areas. XVI. General criteria of moderate seismicity in four regions of the Mediterranean belt.* In *Computational Seismology and Geodynamics* (ed. Chowdhury, D. K.), *vol. 1* (American Geophysical Union, Washington, D.C. 1994) pp. 211–221.

GVISHIANI, A., and SOLOVIEV, A. *On the concentration of major earthquakes around the intersections of morphostructural lineaments in South America.* In *Computational Seismology, vol. 13* (Nauka, Moscow 1980) pp. 46–50 (in Russian).

GVISHIANI, A., GORSHKOV, A., KOSSOBOKOV, V., CISTERNAS, A., PHILIP, H., and WEBER, C. (1987), *Identification of Seismically Dangerous Zones in the Pyrenees*, Annales Geophysicae *5B* (6), 681–690.

GVISHIANI, A., GORSHKOV, A., RANTSMAN, E., CISTERNAS, A., and SOLOVIEV, A., *Identification of Earthquake Prone-areas in Moderate Seismicity Regions* (Nauka, Moscow 1988) (in Russian).

GVISHIANI, A., GORSHKOV, A., KOSSOBOKOV, V., and RANTSMAN, E. (1986), *Morphostructures and Earthquake-prone Areas in the Great Caucasus*, Izvestyia USSR Ac. Sci., Physics of the Earth. *9*, 15–23 (in Russian).

JOO, I. (1992), *Recent Vertical Surface Movements in the Carpathian Basin*, Tectonophysics *202*, 129–134.

KATSKOV, N., and STOICHEV, D. (1976), *Structural-geological Analysis of Satellite Photos of the Balkan Peninsula*, Geologica Balcanica *6* (2), 3–16.

KHAIN, V. E., *Regional Geotectonics: The Alpine-Mediterranean Belt* (Nedra, Moscow 1978) (in Russian).

KOSSOBOKOV V. G., *Recognition of the sites of strong earthquakes by in East Central Asia and Anatolia Hamming's method.* In *Computational Seismology, vol. 14* (Allerton Press, Inc., New York 1983) pp. 78–82.

LINZER, H. G. (1996), *Kinematics of Retreating Subduction along the Carpathian Arc*, Romania Geology *24* (2), 167–170.

LINZER, H. G., FRISCH, W., ZWEIGEL, P., GIRBACEA, R., HANN, H.-P., and MOSER, F. (1998), *Kinematic Evolution of the Romanian Carpathians*, Tectonophysics *297*, 133–156.

MARZA. V., and PANTEA, A., *Seismic hazard in Romania.* In *The Practise of Earthquake Hazard Assessment* (ed. McGuire, R.) (IASPEI and ESC 1993) pp. 213–217.

MCKENZIE, D. P., and MORGAN, W. J. (1969), *The Evolution of Triple Junctions*, Nature *224*, 125–133.

QUANTITATIVE SEISMIC ZONING OF THE CIRCUM-PANNONIAN REGION., *Project CIPA CT 94-0238. First Year Report, March 1996* (eds. Masoli, C., and Panza, G.) (ICTP, Trieste 1996).

QUANTITATIVE SEISMIC ZONING OF THE CIRCUM-PANNONIAN REGION., *Project CIPA CT 94-0238. Second Year Report, March 1997* (eds. Masoli, C., and Panza, G.) (ICTP, Trieste 1997).

PATRASCU, St. (1993), *Paleomagnetic Studies of Some Neogen Magmatic Rocks from the Oas-Ignis-Varatec-Tibles Mountains (Romania)*, Geophys. J. Int. *113*, 215–224.

RADULESCU, D. P., and SANDULESCU, M. (1973), *The Plate Tectonic Concept and the Geological Structure of the Carpathians*, Tectonophysics *16*, 155–161.

RADULIAN, M., MANDRESCU, N., and PANZA, G. (1996a), *Seismogenic Zones of Romania*, ICTP Preprint IC/96/255.

RADULIAN, M., MANDRESCU, N., POPESCU, E., UTALE, A., and PANZA, G. (1996b), *Seismic Activity and Stress Field Characteristics for the Seismogenic Zones of Romania*, ICTP Preprint IC/96/256.

RANTSMAN, E. YA., *Morphostructure of Mountain Regions and Sites of Earthquakes* (Nauka, Moscow 1979) (in Russian).

RATSCHBACHER, L., LINZER, H.-G., MOSER, F., STRUSIEVICZ, R., BEDELEAN, H., HAR, N., and MOGOS, P.-A. (1993), *Cretaceous to Miocene Thrusting and Wrenching along the Central South Carphatians due to a Corner Effect during Collision and Orocline Formation*, Tectonics *12*, 855–873.

REBAI, S., PHILIP, H., and TABOADA, A. (1992), *Modern Tectonic Stress Field in the Mediterranean Region: Evidence for Variation in Stress Directions at Different Scales*, Geophys. J. Int. *110*, 106–140.

RIZNICHENKO, YU. V., *Source dimensions of a shallow earthquake and seismic moment.* In *Research on the Physics of Earthquakes* (Nauka, Moscow 1976) pp. 9–27 (in Russian).

SEISMIC ZONING OF THE USSR TERRITORY (eds. B. Bune, and G. Gorshkov) (Nauka, Moscow 1980).

VELICHKOVA, S., and SOKEROVA, D. (1980), *An Analysis of the Seismic Events in Bulgaria during the 1976*, Bulg. Geophys. J. *6*, 58–72 (in Bulgarian).

(Received April 6, 1998, revised November 25, 1998, accepted April 29, 1999)

Pure appl. geophys. 157 (2000) 97–110
0033–4553/00/020097–14 $ 1.50 + 0.20/0

Pure and Applied Geophysics

Modelling of Block Structure Dynamics for the Vrancea Region: Source Mechanisms of the Synthetic Earthquakes

A. A. SOLOVIEV,[1,2] I. A. VOROBIEVA[1,2] and G. F. PANZA[2,3]

Abstract—The main structural elements (blocks) of the Vrancea (Romania)—the East-European plate; the Moesian, the Black Sea, and the Intra-Alpine (Pannonian-Carpathian) subplates—are assumed to be rigid and separated by infinitely thin plane faults. The interaction of the blocks along the fault planes and with the underlying medium is elasto-viscous. The velocity vectors which define the motion of the East-European plate and of the medium underlying the subplates are input model parameters, which are not changed during the numerical simulation. At each time, the displacements of the subplates caused by these motions are defined so that the system is in the quasistatic equilibrium state. When, for some part of a fault plane, the stress exceeds a certain strength level, a stress-drop (a failure) occurs (in accordance with the dry friction model), and it can cause failures in other parts of the fault planes. The failures represent earthquakes and, as a result of the numerical simulation, a synthetic earthquake catalog is produced. PANZA *et al.* (1997) and SOLOVIEV *et al.* (1999) have determined the ranges for the values of the motion velocity vectors and of the other parameters of the model for which the space distribution of epicenters and the frequency-magnitude relation in the synthetic earthquake catalog are close to those of the observed seismicity. In this paper we study the variation of the slip angle of the synthetic earthquakes as a function of the variation of the model parameters, and we compare our results with observations.

Key words: Block structure dynamics, Vrancea, numerical simulation, source mechanisms.

Introduction

Following ARINEI (1974) the main structural elements of the Vrancea region are: (1) the East-European plate, (2) the Moesian, (3) the Black Sea and (4) the Intra-Alpine (Pannonian-Carpathian) subplates (Fig. 1). The dip angle of the fault system which separates the East-European plate from the Intra-Alpine and Black Sea subplates, and the fault system between the Intra-Alpine and Black Sea subplates is significantly different from 90° (MOCANU, 1993). The main directions of the relative movement of the plates are shown in Figure 1.

---

[1] International Institute of Earthquake Prediction Theory and Mathematical Geophysics, Russian Academy of Sciences, Moscow, Russia.

[2] The Abdus Salam International Centre for Theoretical Physics, SAND Group, Trieste, Italy.

[3] Dipartimento di Scienze della Terra, Universita' di Trieste, Italy.

We simulate the dynamics of the block structure of the region by using the basic principles developed by GABRIELOV *et al.* (1990), SOLOVIEV (1995), PANZA *et al.* (1997), and GORSHKOV *et al.* (1997). Following the assumption that the lithosphere blocks are separated by comparatively thin, weak, less consolidated fault zones such as lineaments and tectonic faults, and that major deformation and most earth-quakes occur in such fault zones, a seismically active region is considered as a system of absolutely rigid blocks divided by infinitely thin plane faults. The blocks interact between themselves and with the underlying medium. The system of blocks moves as a consequence of the prescribed motion of the boundary blocks and of the underlying medium. The motion of the blocks of the structure is defined so that the system is in the quasistatic equilibrium state.

The interaction of the blocks along the fault planes is elasto-viscous (normal state) while the ratio of the stress to the pressure is below a certain strength level. When the critical level is exceeded in some part of a fault plane, a stress-drop (a failure) occurs (in accordance with the dry friction model), and it can cause a failure

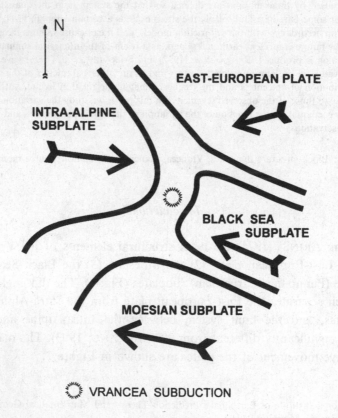

Figure 1
Gross kinematic model proposed for the double subduction process in the Vrancea region, modified after MOCANU (1993).

in other parts of the fault planes. These failures produce earthquakes. Immediately after the earthquake and for some time, the corresponding parts of the fault planes are in the creep state. This state differs from the normal state because of the faster growth of the inelastic displacements and lasts until the stress falls below some other threshold. As a result of the numerical simulation, a synthetic earthquake catalog is produced.

In the model, the strains are accumulated in fault zones. This reflects strain accumulation due to deformations of plate boundaries. Certainly we made considerable simplifications, nonetheless if we want to understand the dependence of earthquake flow on main tectonic movements in a region and its lithosphere structure, simplifications are necessary.

The various numerical experiments carried on by SOLOVIEV et al. (1999) demonstrate that the spatial distribution of epicenters, the slope of the frequency-magnitude (FM) plot, the level of seismic activity, the maximum magnitude of occurred events, and the relative activity of the Vrancea subduction zone with respect to the other considered faults, are sensitive to the directions of the velocity vectors fixed to define the motion of the East-European plate, and of the medium underlying the subplates and to the values of the other parameters of the model. As a result, the motions and the values of the parameters have been found that allow us to obtain a synthetic earthquake catalog with the space distribution of epicenters close to the distribution of the observed seismicity, and with the FM relation, Gutenberg-Richter curve, somehow similar to that obtained for the real earthquake catalog of the Vrancea region.

In this paper we consider the source mechanism, described by the three angles: strike, dip, and slip, of the synthetic earthquakes. Strike and dip define the azimuth and the dip angle of the rupture plane, while the slip defines the direction of the displacement in the rupture plane. Therefore, in our model, strike and dip are prescribed by the block structure geometry and do not depend on the variation of the model parameters. Thus, for the synthetic earthquakes, only the dependence of the slip on the variation of the model parameters is studied, and a comparison is made with observations.

1. Description of the Model

We represent the block structure of the Vrancea region as a limited part of a layer with thickness $H = 200$ km bounded by two horizontal planes (Fig. 2). This thickness corresponds to the depth of the deepest earthquakes in the Vrancea region. The point with geographic coordinates 44.2°N and 26.1°E is chosen as the origin of the reference coordinate system. The X axis is the east-oriented parallel passing through the origin of the coordinate system, the Y axis is the north-oriented meridian passing through the origin of the coordinate system. The lateral

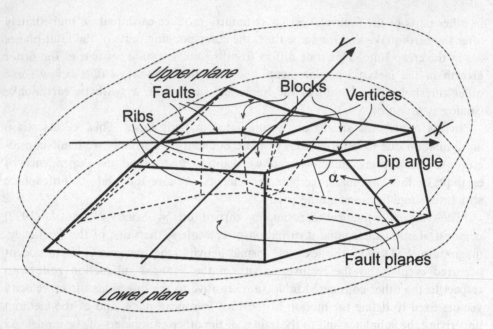

Figure 2
Definitions used in the block structure model.

boundaries of the block structure and its subdivision into blocks are formed by the portions of planes intersecting the layer called "fault planes." The intersection lines of the fault planes with the upper plane are called "the faults." The configuration of the faults of the block structure on the upper plane is presented in Figure 3. The dip angle of each fault plane is given in Table 1.

The movements of the boundaries of the block structure (the boundary blocks) and the medium underlying the blocks generate external forces that act on the block structure. These movements are assumed to be translations, and their rates are considered to be horizontal and known.

Elastic forces arise in the lower plane and in the fault planes as a result of the displacement of the blocks relative to the underlying medium, to the lateral boundary, and to the other blocks. The elastic stress (the force per unit area) at a point is proportional to the difference between the relative displacement and the slippage (the inelastic displacement) at the point. The rate of the inelastic displacement is proportional to the elastic stress. Accordingly,

$$\mathbf{f} = K(\Delta \mathbf{r} - \delta \mathbf{r}), \quad \frac{d\delta \mathbf{r}}{dt} = W\mathbf{f}, \tag{1}$$

where $\mathbf{f}$ is the vector of the elastic stress at the point of the lower plane or of the fault plane, $\Delta \mathbf{r}$ is the vector representing the relative displacement, and $\delta \mathbf{r}$ is the vector representing the inelastic displacement. If the fault plane is considered, then

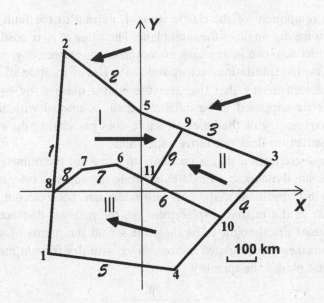

Figure 3
Block structure used in the numerical simulation; the numbers of the vertices (1–11), faults (1–9), and blocks (I–III) are indicated. The arrows outside the block structure indicate the movement of the boundary blocks while those inside the block structure indicate the movement of the underlying medium.

the coefficients K and W in (1) are, respectively, proportional to the shear modulus and inversely proportional to the viscous coefficient of the fault zone.

On the fault plane, the reaction force is normal to the fault plane and its size, per unit area, is:

$$|p_0| = |f_1 \, \text{tg} \, \alpha| \tag{2}$$

Table 1

Basic variant: parameters of faults

Fault #	Vertices of fault	Dip angle	K, bars/cm	W, cm/bars	W_s, cm/bars	Levels of κ		
						B	H_f	H_s
1	1, 8, 2	45°	0	0	0	0.1	0.085	0.07
2	2, 5	120°	1	0.5	1	0.1	0.085	0.07
3	5, 9, 3	120°	1	0.05	0.1	0.1	0.085	0.07
4	3, 10, 4	45°	0	0	0	0.1	0.085	0.07
5	4, 1	45°	0	0	0	0.1	0.085	0.07
6	10, 11, 6	100°	1	0.05	0.1	0.1	0.085	0.07
7	6, 7	100°	1	0.05	0.1	0.1	0.085	0.07
8	7, 8	100°	1	0.05	0.1	0.1	0.085	0.07
9	11, 9	70°	1	0.02	0.04	0.1	0.085	0.07

where f_1 is a component of the elastic stress, $\mathbf{f}$, normal to the fault on the upper plane, and α is the dip angle of the fault plane. The value of p_0 is positive in the case of extension and negative in the case of compression, respectively.

At each time the translation vectors and the angles of rotation of the blocks are determined in such a way that the structure is in a quasi-static equilibrium. All displacements are supposed to be infinitely small, compared with the block size. Therefore the geometry of the block structure does not change during the simulation and the structure does not move as a whole.

The space discretization that is necessary to carry out the numerical simulation of block structure dynamics, is made by splitting the surfaces on which the forces act into cells of trapezoidal shape and with linear size not exceeding 7.5 km. The coordinates X, Y, the relative displacement $\Delta\mathbf{r}$, the inelastic displacement $\delta\mathbf{r}$, and the elastic stress $\mathbf{f}$ are thought to be the same for all the points of a cell.

The earthquakes are simulated in accordance with dry friction model. For each cell of the fault planes the quantity

$$\kappa = \frac{|\mathbf{f}|}{P - p_0} \tag{3}$$

is introduced, where $\mathbf{f}$ is the elastic stress given by (1), P is a parameter of the model which is assumed to be equal for all the faults. P can be interpreted as the difference between the lithostatic (due to gravity) and the hydrostatic pressure, which is assumed to be equal to 2 Kbars for all the faults, and p_0 is the reaction force per unit area, given by (2).

Three values of κ, $B > H_f \geq H_s$, are specified for each fault. It is assumed that the initial conditions, the translation vectors and the angles of rotation of the blocks and the inelastic displacements of the cells, for the numerical simulation satisfy the inequality $\kappa < B$ for all cells of the fault planes. If at any time in one or more cells the value of κ reaches or exceeds the level B, a failure ("earthquake") occurs.

The failure is defined as an abrupt change of the inelastic displacement $\delta\mathbf{r}^e$—in the cell. The new—after the failure—vector of the inelastic displacement $\delta\mathbf{r}^e$ is calculated from

$$\delta\mathbf{r}^e = \delta\mathbf{r} + \delta\mathbf{u}, \quad \delta\mathbf{u} = \gamma\mathbf{f} \tag{4}$$

where $\delta\mathbf{r}$ and $\mathbf{f}$ are the inelastic displacement and the elastic stress, defined by (1), just before the failure and the coefficient γ is determined from the condition that the value of the κ, after the failure, is reduced to the level H_f.

Once the new values of the inelastic displacements for all the failed cells are computed, the translation vectors and the angles of rotation of the blocks are determined to satisfy the condition of the quasi-static equilibrium. If after these computations, for some cell(s) of the fault planes still $\kappa > B$, the procedure is repeated for this (these) cell(s), otherwise the numerical simulation is continued in the ordinary way.

Table 2

Basic variant: parameters of blocks

Block #	Vertices of block	K, bars/cm	W, cm/bars	V_x, cm	V_y, cm
1	2, 8, 7, 6, 11, 9, 5	1	0.05	25	0
2	3, 9, 11, 10	1	0.05	−15	7
3	4, 10, 11, 6, 7, 8, 1	1	0.05	−20	5

On the same fault plane, the cells in which failure occurs at the same time form a single earthquake. The coordinates of the earthquake epicenter are determined as the weighted sum, with weights proportional to the areas of the failed cells, of the coordinates of the cells forming the earthquake. The magnitude of the earthquake is calculated from UTSU and SEKI (1954):

$$M = 0.98 \log_{10} S + 3.93 \tag{5}$$

where S is the total area of the cells forming the earthquake, measured in km^2.

For each earthquake, the source mechanism can be determined considering the vector ΔU, defined as the weighted sum, with weights proportional to the areas of the failed cells, of the vectors δu, given by (4), for the cells forming the earthquake. From (4) and from the definition of f, it follows that ΔU lies in the fault plane where the earthquake occurs.

Immediately after the earthquake, it is assumed that the cells in which the failure occurred are in the creep state. It means that, for these cells, in equation (1), which describes the evolution of the inelastic displacement, the parameter $W_s(W_s > W)$ is used instead of W. After the earthquake, the cell is in the creep state as long as $\kappa > H_s$, when $\kappa \leq H_s$, the cell returns to the normal state and henceforth the parameter W is used in (1) for this cell.

2. Numerical Modelling of Block Structure Dynamics for the Vrancea Region: Basic Variant

PANZA et al. (1997) have found the values of the model parameters for which the considered features of the synthetic and observed seismicity are satisfactorily similar. We refer to this set of parameters as the basic variant. The numerical values of K, W, W_s, B, H_f, and H_s, for each fault, and of K and W, for each block, are given in Tables 1 and 2, respectively. Nondimensional time is used in the model, and all variables containing time are referred to one unit of the nondimensional time. Since $K = 0$ for the boundary faults 1, 4, and 5, in accordance with (1), the relative displacements between the structure blocks and the boundary blocks do not produce forces in the planes of faults 1, 4, and 5.

The components (V_x, V_y) of the velocities of the movement of the underlying medium are given in Table 2 and shown in Figure 3. Boundary faults 2 and 3 move with the same velocity: $V_x = -16$ cm, $V_y = -5$ cm. The directions of the velocities are in accordance with the directions of the main movements of the Vrancea region (Fig. 1).

Quasi-stabilization of the synthetic seismic flow is achieved after 60 unit of the nondimensional time (see PANZA *et al.*, 1997), therefore we consider only the stable part of the synthetic catalog, between the 60th and 150th unit of the nondimensional time.

The synthetic catalog contains 5439 events with magnitude between 5.0 and 7.43. The minimum value of magnitude corresponds, in accordance with (5), to the minimum, area of one cell, while the maximum value (7.43) is close to the value (7.4) observed in reality (MOLDOVEANU *et al.*, 1995).

The FM plots for the observed seismicity of Vrancea and for the synthetic catalog are presented in Figure 4. The curve describing the observed seismicity (solid line) has approximately the same slope (1.33 ± 0.03) as the curve obtained from the synthetic catalog (dashed line).

A considerable part of the synthesized events (49%) occurs on fault 9, and all strong earthquakes $(M > 6.7)$ of the synthetic catalog are concentrated here: the same phenomenon is observed in the real seismicity.

The map with the distribution of the epicenters of the twenty largest synthetic earthquakes (with $M \geq 7.0$) as well as the epicenters of the largest observed Vrancea earthquakes, listed in Table 3, is given in Figure 5.

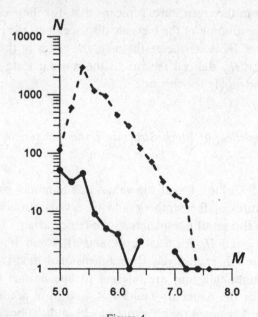

Figure 4

FM plots for the real (solid line) and the synthetic (dashed line) catalogs.

Table 3

Source mechanism of the largest Vrancea earthquakes, as given in the Harvard catalogue (CMT, 1994)

Date	Epicenter		Mechanism, degrees		
	Latitude	Longitude	Strike	Dip	Slip
1977/03/04	45.78°N	26.80°E	235	62	92
1986/08/30	45.51°N	26.47°E	240	72	97
1990/05/30	45.83°N	26.74°E	236	63	101

3. Source Mechanisms of the Synthetic Seismicity

The earthquake source mechanism is described by the values of three angles: strike, dip, and slip. Strike and dip determine the azimuth and the dip angle of the rupture plane, while the slip determines the direction of the displacement in the rupture plane. Therefore, in the model, strike and dip are determined by the block structure geometry: for fault 9, which represents the Vrancea subduction zone, the strike is 212° and the dip is 70°. The slip angle for the synthetic earthquakes is determined by the direction of the vector ΔU, defined in section 1.

The largest Vrancea earthquakes (see Table 3) have similar source mechanisms (CMT, 1994), and the average value of dip ($67° \pm 5°$) corresponds well to the one (70°) of fault 9, while the average strike ($237° \pm 3°$) differs slightly from the one used in the model (212°).

For the basic variant of the model, the slip for the twenty largest synthetic earthquakes, which occurred on the 9th fault, reaches nearly the same value of $99° \pm 1°$, almost equal to the average observed one ($97° \pm 5°$).

To study the dependence of the slip of the synthetic earthquakes on the model parameters, we consider the same variation of parameters as in SOLOVIEV *et al.* (1999): the directions of the velocity vector of the movement of boundary faults 2 and 3, the directions of the velocity vector of the movement of the medium underlying the blocks of the model structure, and the coefficients of the rate of the inelastic displacement, W, on the 2nd and 9th faults. As for the basic variant, we determine in this parametric test the source mechanisms for the twenty largest synthetic earthquakes (the magnitude cutoff is given in Table 4) which occur on the 9th fault.

The values of the slip of the synthetic earthquakes and their standard deviations, obtained for the boundary values of the ranges of the parameter variations, are given in Table 4. The angles given in the second column of Table 4 are the angles between the velocity vector used in the basic variant and the velocity vector used in the parametric test (positive direction is anticlockwise).

In general, and with the exception represented by the variation of the motion direction of the 2nd boundary block, the largest earthquakes on the 9th fault have similar slip (the value of the standard deviation is small), and its value is close to that obtained with the basic variant. The strong variation of the synthetic mechanisms with the variation of the direction of the velocity vector of the 2nd boundary block can be explained by the large variation (140°) of this parameter.

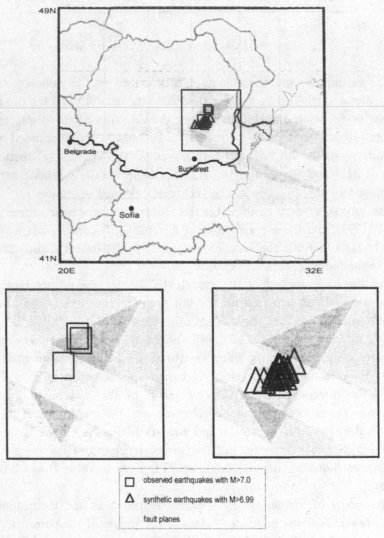

Figure 5

Map the epicenters of the strongest synthetic earthquakes (triangles) and of the strongest observed earthquakes (squares) in the Vrancea region. The projections, on the upper plane, of the fault planes, with $K \neq 0$, are shown as shaded areas.

Table 4

Dependence of the slip angle of the synthetic earthquakes on the variations of the model parameters

Varied parameter		Magnitude cutoff	Slip, degrees
	Basic variant	7.0	99 ± 1
Direction of the movement	1st boundary block, $-10°$	6.9	102 ± 3
	1st boundary block, $+50°$	6.9	99 ± 1
	2nd boundary block, $-140°$	6.0	-70 ± 74
	2nd boundary block, $+70°$	5.6	94 ± 6
	Medium underlying the 1st block, $-50°$	5.8	94 ± 1
	Medium underlying the 1st block, $+50°$	6.3	102 ± 4
	Medium underlying the 2nd block, $-50°$	6.5	101 ± 3
	Medium underlying the 2nd block, $+50°$	6.6	101 ± 2
	Medium underlying the 3rd block, $-50°$	5.0	96 ± 0.1
	Medium underlying the 3rd block, $+50°$	7.0	96 ± 6
Coefficient W	2nd fault, 0.05	5.8	89 ± 6
	2nd fault, 0.7	7.0	99 ± 1
	9th fault, 0.01	6.8	99 ± 1
	9th fault, 0.05	5.9	98 ± 2

The values of slip and their standard deviations for the different directions of the velocity vector of the 2nd boundary block are given in Table 5. The slip does not change when the velocity vector rotates by less than $-60°$. The dispersion of the slip values increases for larger rotations, and earthquakes with different mechanisms occur, though most of the earthquakes are of the over-thrust type, as in the basic variant. When the velocity vector rotates by $-140°$, most of the earthquakes have an extension mechanism.

Table 5

Dependence of the slip angle of the synthetic earthquakes on the rotation of the velocity vector of the 2nd boundary block

Rotation of the velocity vector	Magnitude cutoff	Slip, degrees
0° (basic variant)	7.0	99 ± 1
$-40°$	7.0	98 ± 2
$-60°$	6.6	102 ± 3
$-80°$	6.2	81 ± 58
$-100°$	6.1	-30 ± 90
$-140°$	6.0	-70 ± 74

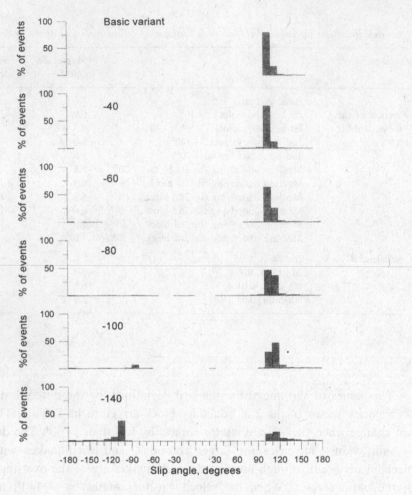

Figure 6
Distribution of the number of synthetic earthquakes versus the slip angle, for the different directions of
the movement of the 2nd boundary block.

In Figure 6, for the different directions of the velocity vector of the 2nd boundary block, the distribution of the number of the synthetic earthquakes versus the slip angle is shown. In the basic variant most earthquakes are of over-thrust type. When the velocity vector rotates clockwise extension earthquakes appear, and two maxima are seen in the lower bar-chart on Figure 6: the larger corresponds to extension earthquakes, and the smaller to over-thrust earthquakes. Thus the modelling of the block structure dynamics shows that, on the same fault, earthquakes with strongly different mechanisms may occur.

Conclusion

The study of the source mechanisms of synthetic earthquakes, obtained with the basic variant of the Vrancea block model described in PANZA *et al.* (1997), shows that earthquakes on the 9th fault have very similar slip angles, which are close to the average values measured for earthquakes occurring in the Vrancea subduction zone. Depending upon some of the velocities of the tectonic movements in the model, it is possible that earthquakes with a different type of source mechanism (different values of the slip) occur along the same fault.

The numerical experiments described in SOLOVIEV *et al.* (1999) show that the direction of the velocity vector of the western part of the East-European plate (the 1st boundary block) and of the Black Sea subplate (the 2nd boundary block) have little influence on the main features of the synthetic catalog. On the contrary, changes in the directions of the motion of the Intra-Alpine and Moesian subplates have the strongest influence on all the features of the synthetic seismicity.

The slip angle is quite stable with respect to the variations of the model parameters, even when the parameters vary within a quite wide range. This is not a surprising result since the force balance changes inconsiderably when one parameter is varied.

Acknowledgements

This work has partly been conducted at the Abdus Salam International Center for Theoretical Physics (Trieste, Italy), supported by INTAS (grant 93-457 extension) and by Russian Foundation for Basic Research (grants 96-05-65710 and 97-05-65802). Support from MURST (40% and 60% funds) has been essential. This is a contribution to UNESCO-IGCP Project 414.

IUGS
UNESCO

Realistic Modelling of Seismic Input for Megacities and Large Urban Areas (project 414)

REFERENCES

ARINEI, St., *The Romanian Territory and Plate Tectonics* (Technical Publishing House, Bucharest 1974) (in Romanian).

CMT (1994), *Centroid Moment Tensor Catalogue*, 1977–1994, Harvard University, Department of Earth and Planetary Sciences, Cambridge, Mass.

GABRIELOV, A. M., LEVSHINA, T. A., and ROTWAIN, I. M. (1990), *Block Model of Earthquake Sequence*, Phys. Earth and Planet. Inter. *61*, 18–28.

GORSHKOV, A. I., KEILIS-BOROK V. I., ROTWAIN, I. M., SOLOVIEV, A. A., and VOROBIEVA, I. A. (1997), *On Dynamics of Seismicity Simulated by the Models of Block-and-faults Systems*, Annali di Geofisica *XL* 5, 1217–1232.

MOCANU, V. I. (1993), Final report "Go West" Programme. Proposal No: 4609. Contract No: CIPA3510PL924609. Subject: *Methods for Investigation of Different Kinds of Lithospheric Plates in Europe*. Period September–December 1993 (Scientific Supervisor: Prof. G. F. Panza).

MOLDOVEANU, C. L., NOVIKOVA, O. V., VOROBIEVA, I. A., and POPA, M. (1995), *The Updated Vrancea Seismoactive Region Catalog*, ICTP, Internal Report, IC/95/104.

PANZA, G. F., SOVOVIEV, A. A., and VOROBIEVA, I. A. (1997), *Numerical Modelling of Block Structure Dynamics: Application to the Vrancea Region*, Pure appl. geophys. *149*, 313–336.

SOLOVIEV, A. A. (1995), *Modelling of Block Structure Dynamics and Seismicity*, European Seismological Commission, XXIV General Assembly, 1994 September 19–24, Athens, Greece, Proceedings and activity Report 1993–1994, Vol. III: University of Athens, Faculty of Sciences, Subfaculty of Geosciences, Department of Geophysics and Geothermy, 1258–1267.

SOLOVIEV, A. A., VOROBIEVA, I. A., and PANZA, G. F. (1999), *Modelling of Block-structure Dynamics: Parametric Study for Vrancea*, Pure appl. geophys. *156*, 395–420.

UTSU, T., and SEKI, A. (1954), *A Relation between the Area of Aftershock Region and the Energy of Main Shock*, J. Seismol. Soc. Japan 7, 233–240.

(Received April 6, 1998, revised October 1, 1998, accepted April 28, 1999)

To access this journal online:
http://www.birkhauser.ch

Pure appl. geophys. 157 (2000) 111–130
0033–4553/00/020111–20 $ 1.50 + 0.20/0

⎡ **Pure and Applied Geophysics**

Stress in the Descending Relic Slab beneath the Vrancea Region, Romania

A. T. ISMAIL-ZADEH,[1,3] G. F. PANZA[2,3] and B. M. NAIMARK[1]

Abstract—We examine the effects of viscous flow, phase transition, and dehydration on the stress field of a relic slab to explain the intermediate-depth seismic activity in the Vrancea region. A 2-D finite-element model of a slab gravitationally sinking in the mantle predicts (1) downward extension in the slab as inferred from the stress axes of earthquakes, (2) the maximum stress occurring in the depth range of 70 km to 160 km, and (3) a very narrow area of the maximum stress. The depth distribution of the annual average seismic energy released in earthquakes has a shape similar to that of the depth distribution of the stress in the slab. Estimations of the cumulative annual seismic moment observed and associated with the volume change due to the basalt-eclogite phase changes in the oceanic slab indicate that a pure phase-transition model cannot solely explain the intermediate-depth earthquakes in the region. We consider that one of the realistic mechanisms for triggering these events in the Vrancea slab can be the dehydration of rocks which makes fluid-assisted faulting possible.

Key words: Stress, slab, Vrancea, numerical modelling.

Introduction

The earthquake-prone Vrancea region is situated at a bend of the Eastern Carpathians and bounded on the north and northeast by the Eastern European platform, on the east and south by the Moesian platform, and on the west by the Transylvanian and Pannonian basins (Fig. 1). The epicenters of mantle earthquakes in the Vrancea region are concentrated within a very small area (less than $1° \times 1°$, Fig. 2a), and the distribution of the epicenters is much denser than that of intermediate-depth events in other intracontinental regions. The projection of the foci on the NW-SE vertical plane across the bend of the Eastern Carpathians (Fig. 2b) shows a seismogenic body in the form of a parallelepiped about 100 km long, about 40-km wide, and extending to a depth of about 180 km. Beyond this depth the seismicity ends suddenly: A seismic event represents an exception beneath 180 km (TRIFU, 1990; TRIFU, *et al.*, 1991; ONCESCU and BONJER, 1997).

---

[1] International Institute of Earthquake Prediction Theory and Mathematical Geophysics, Russian Academy of Sciences, Moscow, Russia.

[2] Dipartimento di Scienze della Terra, Università degli Studi di Trieste, Italy.

[3] Abdus Salam International Center for Theoretical Physics, SAND Group, Trieste, Italy.

As early as 1949, GUTENBERG and RICHTER (1954) drew attention to the remarkable source of shocks in the depth range of 100 km to 150 km in the Vrancea region. According to a historical catalogue (Table 1), there have been 16 large intermediate-depth shocks with magnitudes $M_s > 6.5$ occurring three to five times per century (KONDORSKAYA and SHEBALIN, 1977). In this century, large events in the depth range of 70 to 170 km occurred in 1940 with moment magnitude $M_w = 7.7$, in 1977 $M_w = 7.4$, in 1986 $M_w = 7.1$, and in 1990 $M_w = 6.9$ (ONCESCU and BONJER, 1997).

Using numerous fault-plane solutions for intermediate-depth shocks, RADU (1967), NIKOLAEV and SHCHYUKIN (1975), and ONCESCU and TRIFU (1987) show that the compressional axes are almost horizontal and directed SE-NW, and that the tensional axes are nearly vertical, suggesting that the slip is caused by gravitational forces.

There are several geodynamic models for the Vrancea region (e.g., MCKENZIE, 1970, 1972; FUCHS *et al.*, 1979; RIZNICHENKO *et al.*, 1980; SHCHYUKIN and DOBREV, 1980; CONSTANTINESCU and ENESCU, 1984; ONCESCU 1984; ONCESCU *et al.*, 1984; TRIFU and RADULIAN, 1989; KHAIN and LOBKOVSKY, 1994; LINZER, 1996). MCKENZIE (1970, 1972) suggested that large events in the Vrancea region occur in a vertical relic slab sinking within the mantle and now overlain by continental crust. He believed that the origin of this slab is the rapid southeast motion of the plate containing the Carpathians and the surrounding regions toward the Black Sea plate. The overriding plate pushing from the northwest has formed

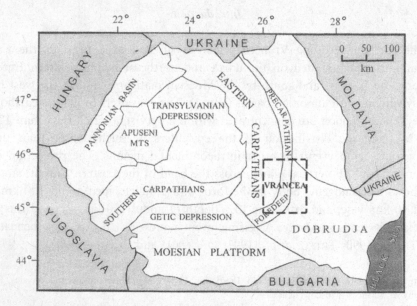

Figure 1
Tectonic sketch of the Carpathian area (modified after RĂDULESCU *et al.*, 1996).

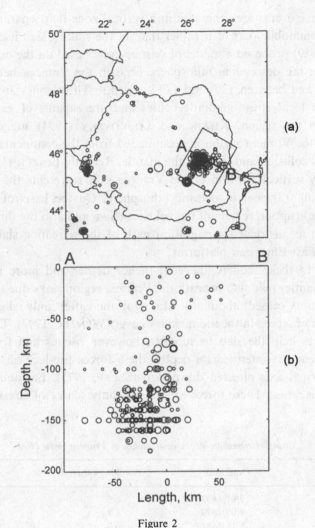

Figure 2
Map of the observed seismicity in Vrancea. (a) Epicenters of Romanian earthquakes with magnitude greater than 4 which occurred from 1900 to 1996. (b) Hypocenters of the same Romanian earthquakes projected onto the vertical plane AB along the NW-SE direction. Several catalogs have been combined to prepare the figure (VOROBIEVA et al., 1996).

the Carpathian orogen, whereas the plate dipping from southeast has evolved the Pre-Carpathian foredeep (RIZNICHENKO et al., 1980). SHCHYUKIN and DOBREV (1980) suggested that the mantle earthquakes in the Vrancea region are to be related to a deep-seated fault descending steeply. The Vrancea region was also considered (FUCHS et al., 1979) as a place where an oceanic slab detached from the continental crust is sinking gravitationally. ONCESCU (1984) and ONCESCU et al. (1984) proposed a double subduction model for Vrancea on the basis of the interpretation of a 3-D seismic tomographic image. In their opinion, the intermediate-depth seismic events are generated in a vertical zone that separates the sinking slab from the immobile part of it rather than in the sinking slab itself. TRIFU and RADULIAN (1989) proposed a model of seismic cycle based on the existence of two active zones in the descending lithosphere beneath the Vrancea between 80- and 110-km depth and between 120- and 170-km depth. These zones are marked by a distribution of local stress inhomogeneities and are capable of generating large earthquakes in the region. KHAIN and LOBKOVSKY (1994) suggested that the lithosphere in the Vrancea region is delaminated from the continental crust during the continental collision and sinks in the mantle. Recently LINZER (1996) proposed that the nearly vertical position of the Vrancea slab represents the final rollback stage of a small fragment of oceanic lithosphere. On the basis of the ages and locations of the eruption centers of the volcanic chain and also the thrust directions, LINZER (1996) reconstructed a migration path of the retreating slab between the Moesian and East-European platforms.

According to these models, the cold (hence denser and more rigid than the surrounding mantle) relic slab beneath the Vrancea region sinks due to gravity. The active subduction ceased about 10 Ma ago; thereafter only slight horizontal shortening was observed in the sedimentary cover (WENZEL, 1997). The hydrostatic buoyancy forces help the slab to subduct, however viscous and frictional forces resist the descent. At intermediate depths these forces produce an internal stress with one principal axis directed downward (SLEEP, 1975). Earthquakes occur in response to this stress. These forces are not the only source of stress that leads to

Table 1

Strong intermediate-depth earthquakes in Vrancea since 1600

No.	Date m/d/y	Magnitude M_s
1	9/01/1637	6.6
2	9/09/1679	6.8
3	8/18/1681	6.7
4	6/12/1701	6.9
5	10/11/1711	6.7
6	6/11/1738	7.0
7	4/06/1790	6.9
8	10/26/1802	7.4
9	11/17/1821	6.7
10	11/26/1829	6.9
11	1/23/1838	6.9
12	10/06/1908	6.8
13	11/01/1929	6.6
14	3/29/1934	6.9
15	11/10/1940	7.4
16	3/04/1977	7.2
17	8/30/1986	6.9
18	5/31/1990	6.7

seismic activity in Vrancea; the process of slab descent may cause the seismogenic stress by means of mineralogical phase changes and dehydration of rocks, which possibly leads to fluid-assisted faulting.

The purpose of this paper is: (1) to study a numerical model of the descending relic slab in an attempt to explain the observed distribution of earthquakes; (2) to examine the influence of the basalt-eclogite phase transition within the slab on the stress in the surrounding rocks; and (3) to discuss a possible role of the dehydration of rocks on the stress release within the descending Vrancea slab.

Viscous Stress in the Descending Slab

Introduction to the Model

Numerical models of subducting slabs have been intensively studied by VASSIL-IOU *et al.* (1984) and VASSILIOU and HAGER (1988) to explain the global depth variation of Benioff zones of seismicity. MAROTTA and SABADINI (1995) showed that the shape of a slab sinking due to its own weight alone differs substantially from the shape of a slab pushed by active convergence. Here, to study the stress distribution and mantle flows beneath the Vrancea region, we construct a model of the evolution of a relic oceanic slab sinking gravitationally beneath an intraconti-nental region.

We assume that, keeping all the other parameters fixed, the number of earth-quakes occurring in Vrancea at intermediate depths is related to the level of viscous stress in the slab. We consider a simple model for the relic slab evolution and calculate the stress therein, assuming that the earth's mantle behaves as a viscous fluid at the geological time scale, and the regional tectonic processes are associated with mantle flows regulated by Newtonian rheology.

The geometry and boundary conditions for the two-dimensional numerical model used in the analysis are shown in Figure 3. A viscous incompressible fluid with variable density and viscosity fills the model square $(0 \leq x \leq L, -H \leq z \leq h)$ divided into four subdomains: atmosphere above $z = 0$, crust, slab, and mantle. These subdomains are bounded by material interfaces where density ρ and viscosity η are discontinuous, but are constant within each subdomain. The interface $z = 0$ approximates a free surface, because the density of the upper layer equals zero, and the viscosity is sufficiently low compared to that in the lower layer. The slab is modeled as being denser than the surrounding mantle, and therefore tends to sink under its own weight.

To test the stability of our results to variations of the density contrast, we consider the value of 0.7×10^2 kg m^{-3}, based on thermal models of the slab (SCHUBERT *et al.*, 1975) and used in numerical modelling of a subducting slab by VASSILIOU *et al.* (1984), and the value 0.4×10^2 kg m^{-3}, suggested by modelling

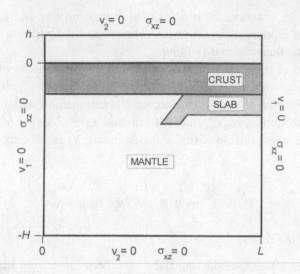

Figure 3
Geometry of the model with the boundary conditions used in the calculations. The z-axis is upward ($z = 0$ approximates the earth's surface), and the x-axis is from left to right.

the long wavelength component of Bouguer anomalies related to the lithospheric roots in the Alps and in the Apennines (WERNER and KISSLING, 1985; MUELLER and PANZA, 1986; MARSON *et al.*, 1995). We also consider several values of the viscosity ratio between the slab and the mantle: 5, 10, and 50, keeping the density contrast equal to 0.4×10^2 kg m^{-3}.

We solve Stokes' equation, which takes the following form in terms of the stream function ψ

$$4 \frac{\partial^2}{\partial x \, \partial z} \eta \frac{\partial^2 \psi}{\partial x \, \partial z} + \left(\frac{\partial^2}{\partial z^2} - \frac{\partial^2}{\partial x^2} \right) \eta \left(\frac{\partial^2 \psi}{\partial z^2} - \frac{\partial^2 \psi}{\partial x^2} \right) = -g \frac{\partial \rho}{\partial x}$$

where g is the acceleration due to gravity; $u = \partial \psi / \partial z$, $v = -\partial \psi / \partial x$, $\mathrm{v} = (u, v)$ is velocity. We assume impenetrability and free-slip boundary conditions:

$$\psi = \partial^2 \psi / \partial x^2 = 0 \quad \text{at} \quad x = 0 \quad \text{and} \quad x = L$$

$$\psi = \partial^2 \psi / \partial z^2 = 0 \quad \text{at} \quad z = -H \quad \text{and} \quad z = h.$$

These boundary conditions keep the model as a closed system, however since the Vrancea oceanic lithosphere is considered as a relic slab sinking in the mantle due only to gravitational forces (MCKENZIE, 1970), we can assume that the external forces are negligible. VASSILIOU *et al.* (1984) studied numerical models of a subducting plate with and without external forces applied to the plate. They showed that minor changes of stress distribution occurred in the plate (and in the system as a whole) due to the forces applied.

The time-dependence of ρ and η is described by the transfer equation

$$\frac{\partial A}{\partial t} = \frac{\partial \psi}{\partial x}\frac{\partial A}{\partial z} - \frac{\partial \psi}{\partial z}\frac{\partial A}{\partial x}$$

where A stands for ρ or η. The position of the material interfaces as functions of time are governed by the following differential equations:

$$dX/dt = \partial\psi/\partial z, \quad dZ/dt = -\partial\psi/\partial x$$

where the points (X, Z) are on the initial interfaces at $t = 0$. The initial distributions $(t = 0)$ of ρ and η and the positions of the material interfaces are known.

To solve the problem, that is, to compute the dependence of density, viscosity, material interfaces, velocity and stress on time, we employ an Eulerian finite element technique described in detail by NAIMARK and ISMAIL-ZADEH (1995), ISMAIL-ZADEH et al. (1996), and NAIMARK et al. (1998). The model region is divided into rectangular elements: 49×47 in the x and z directions. We use dimensionless variables, whereas in presenting the results for stress and velocity we scale them as follows: the time scale t^*, the velocity scale v^*, and the stress scale σ^* are taken respectively as $t^* = \eta^*/[\rho^*g(H+h)]$, $v^* = \rho^*g(H+h)^2/\eta^*$, and $\sigma^* = \rho^*g(H+h)$ where $\eta^* = 10^{20}$ Pa s is a typical value of mantle viscosity (PELTIER, 1984), $\rho^* = 3.3 \times 10^3$ kg m^{-3} is a typical value of mantle density (TURCOTTE and SCHUBERT, 1982).

Numerical Results

The parameter values used in the numerical modelling are listed in Table 2. The deep structure of the crust and uppermost mantle of the Vrancea and surrounding regions is complex. The Moho discontinuity is at about 25–30 km in the basin areas, 30–36 km in the Moesian platform, 38–44 km in the Pre-Carpathian foredeep, and 45–56 km in the Eastern Carpathians (SHCHYUKIN and DOBREV, 1980; ENESCU, 1987). The thickness of the crust in the epicentral region is estimated at 43–44 km from DSS data (RĂDULESCU and POMPILIAN, 1991). As for the underlying mantle, the thickness of the lithosphere varies between less than 100 km within the Carpathian arc and about 150–200 km in the platform areas (SHCHYUKIN and DOBREV, 1980; CHEKUNOV, 1987). In the numerical model we assume that the initial thicknesses of the crust and the slab are 40 km and 30 km, respectively. We choose 45° for the dip of the slab at $t = 0$; changes ($\pm 15°$) in the initial dip of the slab yield results similar to those we describe here. The stress magnitude σ is given by

$$\sigma = [0.5(\tau_{xx}^2 + \tau_{zz}^2 + 2\tau_{xz}^2)]^{1/2} = \eta\left[4\left(\frac{\partial^2\psi}{\partial x\,\partial z}\right)^2 + \left(\frac{\partial^2\psi}{\partial z^2} - \frac{\partial^2\psi}{\partial x^2}\right)^2\right]^{1/2}$$

where τ_{ij} $(i, j = x, z)$ are the components of the deviatoric stress.

The evolution of the slab that sinks under its own weight in the absence of external forces is displayed in Figure 4 for a density contrast 0.7×10^2 kg m^{-3} and

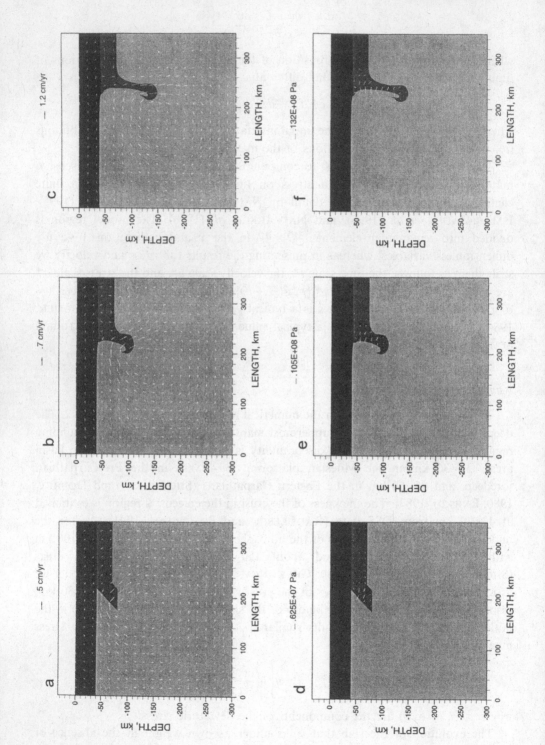

a viscosity ratio 10. The subducting slab gives rise to two mantle flows (Figs. 4a–c). The flow on the left moves clockwise, contributing to the evolution of the Transylvanian basin and the folded arc. The other rotates counterclockwise and possibly affects the development of the Pre-Carpathian foredeep and the Moesian platform. The shape of the slab is controlled by the circulation of mantle material. The mantle flows induced by the slab sinking gravitationally make the slab dip at a higher angle (Fig 4c). Figures 4d–f show the axes of compression of the deviatoric stress. The axes of tension are perpendicular to the axes of compression, and the magnitudes of tension and of compression are the same. The maximum viscous stress is reached within the slab, and the axes of compression are close to the horizontal direction. Based on the JHD method providing most relative locations of hypocenters, TRIFU (1990) and TRIFU et al. (1991) showed that the hypocentral projection of Vrancea intermediate-depth earthquakes onto the vertical plane along the NW-SE direction are nearly vertical and extended downward over the whole depth range. Most recently, ONCESCU and BONJER (1997) relocated the best recorded microearthquakes in the Vrancea region during 1982–1989 and

Table 2

Model parameters

Notation	Meaning	Value
g	acceleration due to gravity, m s^{-2}	9.8
h	height over the surface, km	33
h_c	initial thickness of the crust, km	40
h_s	initial thickness of the slab, km	30
$H+h$	vertical size of the model, km	333
L	horizontal size of the model, km	350
t^*	time scale, yr	300
v^*	velocity scale, m yr^{-1}	1.1×10^3
η^*	typical value of viscosity, Pa s	10^{20}
η_{air}	viscosity over the surface, Pa s	10^{15}
η_c	viscosity of the crust, Pa s	10^{22}
η_m	viscosity of the mantle, Pa s	10^{20}
η_s	viscosity of the slab, Pa s	10^{21}
ρ^*	typical value of density, kg m^{-3}	3.3×10^3
ρ_{air}	density over the surface, kg m^{-3}	0
ρ_c	density of the crust, kg m^{-3}	2.9×10^3
ρ_m	density of the mantle, kg m^{-3}	3.3×10^3
ρ_s	density of the slab, kg m^{-3}	3.37×10^3 and 3.34×10^3
σ^*	stress scale, Pa	1.1×10^{10}

Figure 4

Flow fields (a–c) and deviatoric compression axes (d–f) for the evolution of the slab subject to gravitational forces only: (a, d) $t = 16$ Ma BP. (b, e) $t = 10$ Ma BP, (c, f) present-day. The maximum values of flow velocity and stress magnitude are shown at the top of the figures.

showed (a) again the nearly vertical distribution of hypocenters of the events and (b) very narrow zone of the seismic activity (about 10-km wide). The numerical results, indicating a maximum stress in the narrow and subvertical region of the model, are in agreement with the observations.

The same computations made with a density contrast of 0.4×10^2 kg m^{-3} produce a nearly identical pattern. The numerical results indicate that variations of the viscosity ratio lead to changes in the stress distribution and in the velocity of the descending slab. If the viscosity ratio between the slab and the surrounding mantle is as small as 5, then the stress in the slab is not large enough. A high viscosity ratio (50) causes a slow descent of the slab (about 0.3 cm yr^{-1}), while the stress is now sufficiently large to give rise to seismic activity. Our computations show that a viscosity ratio of 10 is more suitable for the Vrancea region, because in this case the velocity of slab descent is about 1–2 cm yr^{-1}, which agrees with the regional geological inferences (BLEAHU *et al.*, 1973) and with the rate of subduction predicted from the thermal model of lithosphere in the Vrancea (DEMETRESCU and ANDREESCU, 1994).

The depth distribution of the average stress magnitude in the slab for the two density contrasts considered is presented in Figure 5. To compare the stress distribution resulting from the model with regional observations, we plot annual average energy released in earthquakes E versus depth. To do this, we use a combined catalog of earthquakes in the Vrancea region (VOROBIEVA *et al.*, 1996). This catalog consists of the subcatalogs of RADU (1979) for 1932–1979, earthquakes in the USSR from 1962–1990 (computer data file, 1992) for 1962–1979,

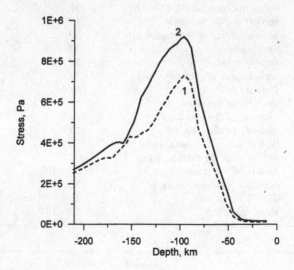

Figure 5
Depth distribution of the average stress in the model for density contrasts 0.4×10^2 kg m^{-3} (1) and 0.7×10^2 kg m^{-3} (2).

TRIFU and RADULIAN (1991) for 1980–1991, and world hypocenter data file USGS-NEIC for 1991–1996. However, we should note that the RADU catalog (1979) contains no depths. Accordingly, we analyze the distribution of earthquakes over magnitude and depth for 1962 to 1996. To calculate annual average energy E, we employ the GUTENBERG and RICHTER (1954) relation between E and magnitude M_s: $\log E = 1.5\ M_s + 11.8$. The two computed curves in Figure 5 show that the stress is the largest in the depth range from about 70 km to 150 km and has a shape similar to that of bar charts of $\log E$ versus depth (Fig. 6). A close inspection of the curves in Figure 5 and of the graph in Figure 6 reveals that the maximum viscous stress is reached at a depth of about 90 km, whereas the maximum energy released by earthquakes is observed at a depth of about 110 km. There is the second peak in energy distribution at a depth of about 150 km. The existence of stress heterogeneities responsible for earthquake occurrence send us to consider other faulting processes at intermediate depths.

Intermediate-depth Faulting Processes

Large earthquakes in the Vrancea region occur within a relic slab sinking in the mantle. It is less obvious that the observed time-space distribution of the large Vrancea events might be explained solely by viscous stress release. High-pressure faulting processes at intermediate depths in the Vrancea slab can also be activated

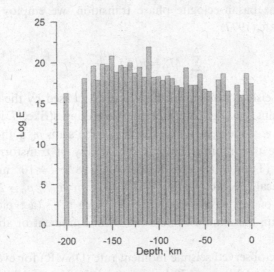

Figure 6
Distribution of the annual average seismic energy released E (measured in J) in 5-km depth intervals in the Vrancea region for the period 1932 to 1996.

by the stress produced by heterogeneities in the volume change due to phase transitions, and/or by the dehydration of rocks, which possibly leads to fluid-assisted faulting.

Phase Transition, Seismic Moment, and Volume Change

Slab metamorphism plays a crucial role in faulting processes at high pressures. Many authors have considered intermediate-depth earthquakes as a result of phase changes from basalt to eclogite in the slab (e.g., COMTE and SUÁRES, 1994). There are two main effects of these exothermic phase transitions (with a small positive Clapeyron slope): deflections of the phase boundary from its normal position and release of latent heat. As for the latter, it slightly changes the temperature of the surrounding material (S. Karato and S. Sobolev, personal communication, 1995) and hence the buoyancy forces. Deflection of the phase boundary depends upon the lateral temperature difference occurring in a relatively cold slab sinking into a hot mantle. The effects of phase transitions in the slab have two implications for the state of stress: (1) the volume change results in a contraction in the direction of the maximum principal stress and in increased compressive stress; and (2) the denser phase acts as an additional load that pulls down the slab and causes an increase of the viscous stress. As a volume within a rock mass undergoes transformation to a denser phase, contraction occurs in the direction of the maximum compressive stress, and large deviatoric stresses are generated within the neighboring rocks, leading to seismic failure.

To estimate the effect on the Vrancea seismicity due to the volume change associated with the basalt-eclogite phase transition, we employ the relation suggested by McGARR (1977)

$$\sum_{n=1}^{N} M_0^n = \mu l T v_s \frac{\rho_1 - \rho_0}{\rho_1}$$

where M_0^n, is the seismic moment of the nth event caused by the volume change, μ is the shear modulus, l is the length of the slab along strike, T is the thickness of the oceanic crust, v_s is the velocity of descent of the slab, ρ_0 is the density of rocks prior to the phase transition, and ρ_1 is the density of transformed rocks. Given $\mu = 6.5 \times 10^{10}$ Pa (TURCOTTE and SCHUBERT, 1982), $l = 10^5$ m, $T = 10^4$ m (the thickness of a typical oceanic crust), $v_s = 2 \times 10^{-2}$ m yr^{-1}, $\rho_0 = 2.92 \times 10^3$ kg m^{-3} (a typical density of wet basalts), $\rho_1 = 3.5 \times 10^3$ kg m^{-3} (a typical density of dry eclogites), we obtain the annual cumulative seismic moment of about 2×10^{17} N m yr^{-1}.

To estimate the observed seismic moment rate (OSMR) for events in Vrancea in the depth range from 60 km to 170 km, we used the Harvard University Centroid-Moment Tensor Catalog (a computer file, 1977–1995). This catalog contains events with $M \geq 5$, and occasionally with lower magnitudes; the eight largest shocks are listed in Table 3. OSMR is found to be about 1.6×10^{19} N m yr^{-1} for the region. We consider a time period of 19 years that includes most of the largest earthquakes occurring in the region during the last century. In the evaluation of OSMR, the time period considered should be long enough to provide a representative sample of large earthquakes in the region. If the time interval is too short and does not include the largest shocks, it can result in an underestimate of OSMR and, conversely, one may overestimate the moment rate, if the time window encloses an unusual sequence of large events.

If we extend the time window to 1900 in the estimation of the annual OSMR, we must include the 1940 earthquake, with $M_w = 7.7$, a focal depth of 150 km and seismic moment, M_0 of 5.1×10^{20} (ONCESCU and BONJER, 1997). Hence, for this century, we get an OSMR of at least 8×10^{18} N m yr^{-1}. This value can be representative of a longer period of time, considering that the large earthquakes that have occurred since 1600 seem to follow a regular pattern (PURCARU, 1979; RIZNICHENKO et al., 1980; NOVIKOVA et al., 1995).

Thus the cumulative annual seismic moment associated with the volume change due to the phase transition is lower than that obtained from observations, so that a pure phase-transition model cannot explain the intermediate-depth seismicity in Vrancea.

Dehydration-induced Faulting

According to the subduction model for the thermal structure of the Eastern Carpathians, the seismogenic Vrancea zone lies above the 800°C isotherm, which approximately marks the brittle/ductile transition for ultramafic materials (DEME-TRESCU and ANDREESCU, 1994). The strength envelop calculated for the Vrancea region points to a strong upper lithosphere where the tensional stress can range up to about 1000 MPa (LANKREIJER et al., 1997).

Table 3

Subcatalog of strong intermediate-depth earthquakes in Vrancea beginning with 1977 event

No.	Date	Time	Latitude	Longitude	Depth	M_0
	m/d/y	h:m:s	°N	°E	km	N m
1	3/04/77	19:21:54	45.77	26.76	84	1.99×10^{20}
2	10/02/78	20:28:53	45.72	26.47	154	4.75×10^{16}
3	5/31/79	07:20:06	45.54	26.32	114	7.26×10^{16}
4	9/11/79	15:36:54	45.56	26.29	143	6.23×10^{16}
5	8/01/85	14:35:03	45.74	26.50	103	7.96×10^{16}
6	8/30/86	21:28:36	45.54	26.29	133	7.91×10^{19}
7	5/30/90	10:40:06	45.86	26.67	74	3.01×10^{19}
8	5/31/90	00:17:48	45.79	26.75	87	3.23×10^{18}

Despite the fact that rocks in the subducting slabs have considerably more strength compared with the surrounding material, the frictional processes resulting from pressure prevent brittle failure. At pressures above 3 GPa (about 100 km of depth), and even at a temperature of 20°C, brittle failure of rock is impossible in the absence of fluids (GREEN and HOUSTON, 1995). On the basis of experimental investigations, RALEIGH and PATERSON (1965) demonstrated that serpentinites (serpentinized peridotites) become brittle as a result of dehydration at high pressures such for which unhydrous rocks are plastically deformed.

It is well known from fracture mechanics that microcracks in rock are generated during brittle failure due to a tensile process (e.g., BRACE and BOMBOLAKIS, 1963; SCHOLZ, 1990; ROTWAIN *et al.*, 1997). The fluid released by dehydration fills the cracks and the pore fluid contributes, together with the stress, to the opening of microcracks by filling them. As macroscopic stress continues to rise, the tensile strength is exceeded and, finally, in some local region the rock becomes fractured so that it loses its ability to support the compressive load, with the resulting formation of a small fault within this region. The fault is bounded by a zone with a high density of tensile microcracks. This zone filled by fluid thus becomes the principal seat of the pore pressure generation that is necessary for fault growth.

Consequently, if a source of volatiles is available, there is a possibility of high-pressure faulting in the slab beneath Vrancea. Obviously, H_2O is carried down with the sediments covering the uppermost part of the slab, and the hydrated oceanic crust contains about 2% of H_2O at 3.0 GPa and 700°C. Moreover, results of recent experimental studies (ULMER and TROMMSDORFF, 1995) show that the subduction of serpentinites containing about 13% of H_2O may transport large quantities of water to depths of the order of 150–200 km. VANYAN (1997) believes that the reaction of dehydration occurring in the sinking slab can easily be detected as zones of electrical conductivity anomalies. In the Vrancea region the electrical resistivity drops below 1 Ω m (STĂNICĂ and STĂNICĂ, 1993) and indicates the upper limit of a conducting zone that correlates with the Carpathian electrical conductivity anomaly (PINNA *et al.*, 1992). Thus, the dehydration-induced faulting in the depth range of 70 to 170 km can contribute to the increase of stress and consequently to the intermediate-depth seismicity observed in Vrancea. This is mainly a qualitative inference; a quantitative estimate of the seismic moment rate associated with the dehydration of minerals in the slab is to be the subject of other specific research.

Discussion and Conclusions

There are essential distinctions between the intermediate-depth seismicity in intracontinental regions and the ordinary Benioff zones. The Pacific seismic zone is a linear extended structure several thousands of km in length and hundreds of km in width. Earthquakes with focal depth up to 60 km dominate these regions. At the same time, the earthquakes in the Circum-Pacific belt clearly concentrate on a nearly continuous circle along the subduction zones, while the seismicity of the Alpine-Himalayan orogenic belt is diffuse and does not correlate with active subduction zones. According to KHAIN and LOBKOVSKY (1994) the intermediate-depth events are observed in southern Spain, Calabria, Hellenic arc, Vrancea, Caucasus, Zagros, Pamir-Hindu Kush, and Assam (Fig. 7).

SPAKMAN (1991) and DE JONGE et al. (1994) used seismic tomography to reveal oceanic slabs sinking beneath the Alboran (southern Spain), Calabrian, and southern Aegean regions. The Vrancea and Pamir-Hindu Kush regions are particularly remarkable in that mantle seismicity is concentrated within very narrow zones in the sinking slabs. The intermediate-depth earthquakes in the Caucasus are likely to be associated with a relic Benioff zone dipping under the Greater Caucasus (KHALILOV et al., 1987; GODZIKOVSKAYA and REYSNER, 1989). Therefore, the distinguishing feature of the Alpine-Himalaya seismic belt is its intermediate-depth events in paleosubducted slabs.

Studying the K_2O/SiO_2 ratio for magmatic rocks, BOCCALETTI et al. (1973) and BLEAHU et al. (1973) suggested that the Vrancea slab was subducted during Neogene time and reached depths of about 160 km where it partially melted and generated calc-alkaline magmas which erupted behind the Carpathian folded arc, building up the magmatic arc. They also believe that the persisting subduction caused an active stretching of the Transylvanian basin and eruption of basaltic magma in the Quaternary. The finite-element model of a descending relic slab allows us to explain the seismic activity in Vrancea: the axes of compression and tension are close to the horizontal and vertical directions, respectively; the maximum viscous stress is found to be at depths of 70 km to 160 km; the model predicts a very narrow area of maximum stress. The simplified numerical model explains, although roughly, the intermediate-depth seismicity in the region, if the seismic energy release depends exponentially on stress. Considering that hypocenters of large earthquakes in Vrancea fall in the narrow area (TRIFU, 1990; TRIFU et al., 1991; ONCESCU and BONJER, 1997), a 3-D modelling of sinking slab seems to be more appropriate for the Vrancea region. Nevertheless the analyzed 2-D model reproduces the main features of spatial distribution of stress in the region.

The seismic moment rate due to the volume change associated with the effect of the basalt-eclogite phase transition in the descending slab is much lower (about 40 times) than that obtained from the events in Vrancea in the depth range of 60 km to 170 km. This suggests that the volume reduction is not likely to significantly contribute to the stress buildup at intermediate depths in Vrancea. Alternatively, the generation of a pore fluid by dehydration of hydrous minerals in the slab may give rise to dehydration-induced faulting. Thus, viscous flows due to the sinking relic slab together with the dehydration-induced faulting can be considered as a plausible triggering mechanism explaining the intermediate-depth seismicity in Vrancea.

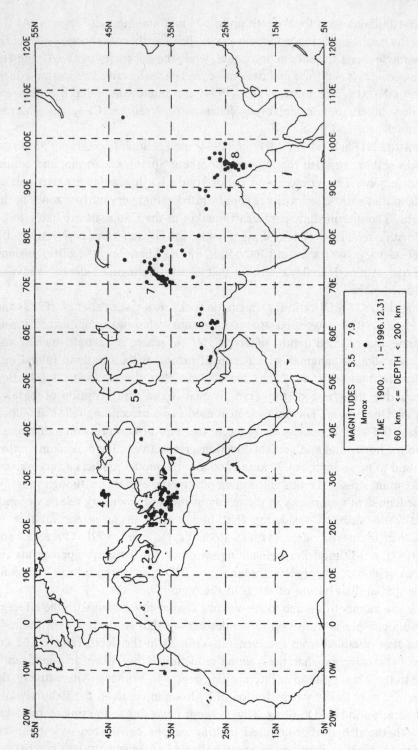

Figure 7

Spatial distribution of intermediate-depth earthquakes with $M_s > 5.5$ in the Alpine-Himalaya seismic belt. (1) Southern Spain; (2) Calabria; (3) the Hellenian arc; (4) Vrancea; (5) Caucasus; (6) Zagros; (7) Pamir-Hindu Kush; (8) Assam.

At the same time, the suggested model and hypotheses of stress generation should not be overestimated, for they still have many limitations and assumptions. The numerical model of a sinking slab cannot explain two separated zones of distinct seismicity in the Vrancea lithosphere at depths of 80 to 110 km and 120 to 170 km as suggested by TRIFU and RADULIAN (1989). The hypothesis of phase changes at the intermediate depths does not support the existing focal mechanism solutions for these earthquakes. The model of stress generation due to dehydration still remains conceptual, because a quantitative estimation of the rate of seismic moment associated with dehydration-induced faulting is required. Hence the model of stress generation in the Vrancea region must be improved to better understand and explain the origin of the intermediate-depth earthquakes.

Acknowledgements

The Harvard University Seismic Centroid-Moment Tensor Catalog was kindly provided via Internet by the group of A. Dziewonski. We are very grateful to V. Keilis-Borok, T. Kronrod, G. Molchan, Yu. Shchyukin, I. Vorobieva and F. Wenzel for useful discussions of this research, to C. Frohlich, W. Jacoby, V. Khain, M. Radulian, C.-I. Trifu, and one anonymous referee for their thorough review of the initial versions of the manuscript and useful comments. We are also thankful to R. Nicolich and L. Cernobori for the computing facilities at DINMA, University of Trieste, and to I. Vorobieva for providing us with the combined catalog of the Vrancea region. This publication was partially supported by NATO (ENVIRLG 931206), by INTAS (93-0809-ext), by the U.S. NSF (EAR-9423818), and by the RFBR (95-05-14083).

REFERENCES

BRACE, W. F., and BOMBOLAKIS, A. G. (1963), *A Note on Brittle Crack Growth in Compression*, J. Geophys. Res. *68*, 3709–3713.

BLEAHU, M. D., BOCCALETTI, M., MANETTI, P., and PELTZ, S. (1973), *Neogene Carpathian Arc: A Continental Arc Displaying the Features of an 'Island Arc'*, J. Geophys. Res. *78*, 5025–5032.

BOCCALETTI, M., MANETTI, P., PECCERILLO, A., and PELTZ, S. (1973), *Young Volcanism in the Calimani-Harghita Mountains (East Carpathians): Evidence of a Paleoseismic Zone*, Tectonophysics *19*, 299–313.

CHEKUNOV, A. V. (1987), *A Tectonic Model of the Seismoactive Region of Vrancea in the Carpathians*, Geologichesky zhurnal *4*, 3–11 (in Russian).

COMTE, D., and SUÁRES, G. (1994), *An Inverted Double Seismic Zone in Chile: Evidence of Phase Transformation in the Subducted Slab*, Science *263*, 212–215.

CONSTANTINESCU, L., and ENESCU, D. (1984), *A Tentative Approach to Possibly Explaining the Occurrence of the Vrancea Earthquakes*, Rev. Roum. Geol. Geophys. Geogr. *28*, 19–32.

DEMETRESCU, C., and ANDREESCU, M. (1994), *On the Thermal Regime of Some Tectonic Units in a Continental Collision Environment in Romania*, Tectonophysics *230*, 265–276.

ENESCU, D. (1987), *Contributions to the Knowledge of the Lithosphere Structure in Romania on the Basis of Seismic Data*, Rev. Roum. Geol. Geophys. Geogr. *31*, 20–27.

FUCHS, K., BONJER, K.-P., BOCK, G., CORNEA, I., RADU, C., ENESCU, D., JIANU, D., NOURESCU, A., MERKLER, G., MOLDOVEANU, T., and TUDORACHE, G. (1979), *The Romanian Earthquake of March 4, 1977. II. Aftershocks and Migration of Seismic Activity*, Tectonophysics *53*, 225–247.

GODZIKOVSKAYA, A. A., and REYSNER, G. I. (1989), *Endogenetic Location of Deep Earthquakes in Caucasus*, Geotectonics *3*, 15–25 (in Russian).

GREEN, H. W. II, and HOUSTON, H. (1995), *The Mechanics of Deep Earthquakes*, Ann. Rev. Earth Planet. Sci. *23*, 169–213.

GUTENBERG, B., and RICHTER, C. F., *Seismicity of the Earth*, 2nd ed. (Princeton University Press, Princeton, N.J. 1954).

ISMAIL-ZADEH, A. T., NAIMARK, B. M., and LOBKOVSKY, L. I., *Hydrodynamic model of sedimentary basin formation based on development and subsequent phase transformation of a magmatic lens in the upper mantle*. In *Computational Seismology and Geodynamics*, *vol. 3* (ed. Chowdhury, D. K.) (Am. Geophys. Un., Washington, D.C. 1996) pp. 42–53.

DE JONGE, M. R., WORTEL, M. J. R., and SPAKMAN, W. (1994), *Regional Scale Tectonic Evolution and the Seismic Velocity Structure of the Lithosphere and Upper Mantle: The Mediterranean Region*, J. Geophys. Res. *99*, 12,091–12,108.

KHAIN, V. E., and LOBKOVSKY, L. I. (1994), *Conditions of Existence of the Residual Mantle Seismicity of the Alpine Belt in Eurasia*, Geotectonics *3*, 12–20 (in Russian).

KHALILOV, E. N., MEKHTIEV, SH. F., and KHAIN, V. E. (1987), *On Geophysical Data Confirming the Collisional Origin of the Greater Caucasus*, Geotectonics *2*, 54–60 (in Russian).

KONDORSKAYA, N. V., and SHEBALIN, N. V. (eds.), *New Catalog of Large Earthquakes in the USSR from Antiquity to 1975* (Nauka, Moscow, 1977).

LANKREIJER, A., MOCANU, V., and CLOETINGH, S. (1997), *Lateral Variations in Lithosphere Strength in the Romanian Carpathians: Constraints on Basin Evolution*, Tectonophysics *272*, 269–290.

LINZER, H.-G. (1996), *Kinematics of Retreating Subduction Along the Carpathian Arc, Romania*, Geology *24* (2), 167–170.

MAROTTA, A. M., and SABADINI, R. (1995), *The Style of the Tyrrhenian Subduction*, Geophys. Res. Lett. *22*, 747–750.

MARSON, I., PANZA, G. F., and SUHADOLC, P. (1995), *Crust and Upper Mantle Models Along the Active Tyrrhenian Rim*, Terra Nova *7*, 348–357.

MCGARR, A. (1977), *Seismic Moment of Earthquakes Beneath Island Arc, Phase Changes, and Subduction Velocities*, J. Geophys. Res. *82*, 256–264.

MCKENZIE, D. P. (1970), *Plate Tectonics of the Mediterranean Region*, Nature *226*, 239–243.

MCKENZIE, D. P. (1972), *Active Tectonics of the Mediterranean Region*, Geophys. J. R. Astron. Soc. *30*, 109–185.

MUELLER, S., and PANZA, G. F. (1986), *Evidence of a Deep-reaching Lithospheric Root Under the Alpine Arc*, Developments in Geotectonics *21*, 93–113.

NAIMARK, B. M., and ISMAIL-ZADEH, A. T. (1995), *Numerical Models of Subsidence Mechanism in Intracratonic Basins: Application to North American Basins*, Geophys. J. Int. *123*, 149–160.

NAIMARK, B. M., ISMAIL-ZADEH, A. T., and JACOBY, W. R. (1998), *Numerical Approach to Problems of Gravitational Instability of Geostructures with Advected Material Boundaries*, Geophys. J. Int. *134*, 473–483.

NIKOLAEV, P. N., and SHCHYUKIN, YU. K., *Model of crust and uppermost mantle deformation for the Vrancea region*. In *Deep Crustal Structure* (Nauka, Moscow 1975) pp. 61–83 (in Russian).

NOVIKOVA, O. V., VOROBIEVA, I. A., ENESCU, D., RADULIAN, M., KUZNETZOV, I., and PANZA, G. F. (1995), *Prediction of the Strong Earthquakes in Vrancea, Romania, Using the CN Algorithm*, Pure appl. geophys. *145*, 277–296.

ONCESCU, M. C. (1984), *Deep Structure of the Vrancea Region, Romania, Inferred from Simultaneous Inversion for Hypocenters and 3-D Velocity Structure*, Ann. Geophys. *2*, 23–28.

ONCESCU, M. C., and BONJER, K. P. (1997), *A Note on the Depth Recurrence and Strain Release of Large Vrancea Earthquakes*, Tectonophysics *272*, 291–302.

ONCESCU, M. C., BURLACU, V., ANGHEL, M., and SMALBERGHER, V. (1984), *Three-dimensional P-wave Velocity Image under the Carpathian Arc*, Tectonophysics *106*, 305–319.

ONCESCU, M. C., and TRIFU, C.-I. (1987), *Depth Variation of Moment Tensor Principal Axes in Vrancea (Romania) Seismic Region*, Ann. Geophys. *5*, 149–154.

PELTIER, W. R. (1984), *The Rheology of the Planetary Interior*, Rheology *28*, 665–697.

PINNA, E., SOARE, A., STĂNICĂ, D., and STĂNICĂ, M. (1992), *Carpathian Conductivity Anomaly and its Relation to Deep Substratum Structure*, Acta Geodaet. Geophys. Mont. Hung. *27*, 35–45.

PURCARU, G. (1979), *The Vrancea, Romania, Earthquake of March 4, 1977—A Quite Successful Prediction*, Phys. Earth Planet. Inter. *18*, 274–287.

RADU, C. (1967), *On the Intermediate Earthquakes in the Vrancea Region*, Rev. Roum. Geol. Geophys. Geogr. *11*, 113–120.

RADU, C., *Catalogue of strong earthquakes originated on the Romanian territory Part II: 1901–1979*. In *Seismological Researches on the Earthquake of March 4, 1977—Monograph* (eds. Cornea, I., and Radu, C.) (Central Institute of Physics, Bucharest 1979).

RĂDULESCU, F., and POMPILIAN, A. (1991), *Twenty-five Years of Deep Seismic Soundings in Romania (1966–1990)*, Rev. Roum. Geol. Geophys. Geogr. *35*, 89–97.

RĂDULESCU, F., MOCANU, V., NACU, V., and DIACONESCU, C. (1996), *Study of Recent Crustal Movements in Romania: A Review*, J. Geodyn. *22*, 33–50.

RALEIGH, C. B., and PATERSON, M. S. (1965), *Experimental Deformation of Serpentine and its Tectonic Consequences*, J. Geophys. Res. *70*, 3965–3985.

RIZNICHENKO, YU. V., DRUMYA, A. V., STEPANENKO, N. YA., and SIMONOVA, N. A., *Seismicity and seismic risk of the Carpathian region*. In *The 1977 Carpathian Earthquake and its Impact* (ed. Drumya, A. V.) (Nauka, Moscow 1980) pp. 46–85 (in Russian).

ROTWAIN, I., KEILIS-BOROK, V., and BOTVINA, L. (1997), *Premonitory Transformation of Steel Fracturing and Seismicity*, Phys. Earth Planet. Inter. *101*, 61–71.

SCHOLZ, C., *The Mechanics of Earthquake and Faulting* (Cambridge University, Cambridge 1990).

SCHUBERT, G., YUEN, D. A., and TURCOTTE, D. L. (1975), *Role of Phase Transitions in a Dynamic Mantle*, Geophys. J. R. Astron. Soc. *42*, 705–735.

SHCHYUKIN, YU. K., and DOBREV, T. D., *Deep geological structure, geodynamics and geophysical fields of the Vrancea region*. In *The 1977 Carpathian Earthquake and its Impact* (ed. Drumya, A. V.) (Nauka, Moscow 1980) pp. 7–40 (in Russian).

SLEEP, N. H. (1975), *Stress and Flow Beneath Island Arcs*, Geophys. J. R. Astron. Soc. *42*, 827–857.

SPAKMAN, W. (1991), *Tomographic Images of the Upper Mantle Below Central Europe and the Mediterranean*, Terra Nova *2*, 542–553.

STĂNICĂ, D., and STĂNICĂ, M. (1993), *An Electrical Resistivity Lithospheric Model in the Carpathian Orogen from Romania*, Phys. Earth Planet. Inter. *81*, 99–105.

TRIFU, C.-I. (1990), *Detailed Configuration of Intermediate Seismicity in the Vrancea Region*, Rev. de Geofisica *46*, 33–40.

TRIFU, C.-I., and RADULIAN, M. (1989), *Asperity Distribution and Percolation as Fundamentals of Earthquake Cycle*, Phys. Earth Planet. Inter. *58*, 277–288.

TRIFU, C.-I., and RADULIAN, M. (1991), *A Depth-magnitude Catalog of Vrancea Intermediate-depth Microearthquakes*, Rev. Roum. Geol. Geophys. Geogr. *35*, 35–45.

TRIFU, C.-I., DESCHAMPS, A., RADULIAN, M., and LYON-CAEN, H. (1991), *The Vrancea earthquake of May 30, 1990: An estimate of the source parameters*, Proceedings of the XXII General Assembly of the European Seismological Commission, Barcelona, 449–454.

TURCOTTE, D. L., and SCHUBERT, G., *Geodynamics: Application of Continuum Physics to Geological Problems* (John Wiley & Sons, New York 1982).

ULMER, P., and TROMMSDORFF, V. (1995), *Serpentine Stability to Mantle Depths and Subduction-related Magmatism*, Science *268*, 858–861.

VANYAN, L. L. (1997), *Electromagnetic Evidence for Fluid in the Crust and Upper Mantle East and West of the TTZ*, Ann. Geophys. *15*, Suppl. *I*. C159.

VASSILIOU, M. S., and HAGER, B. H. (1988), *Subduction Zone Earthquakes and Stress in Slabs*, Pure appl. geophys. *128*, 547–624.

VASSILIOU, M. S., HAGER, B. H., and RAEFSKY, A. (1984), *The Distribution of Earthquakes with Depth and Stress in Subducting Slabs*, J. Geodyn. *1*, 11–28.

VOROBIEVA, I. A., NOVIKOVA, O. V., KUZNETSOV, I. V., ENESCU, D., RADULIAN, M., and PANZA, G. (1996), *Intermediate-term Earthquake Prediction for the Vrancea Region: Analysis of New Data*, Computational Seismology *28*, 83–99 (in Russian).
WENZEL, F. (1997), *Tectonics of the Vrancea zone (Romania).* In *Abstracts, 29th General Assembly of IASPEI* (Thessaloniki, Greece 1997) p. 280.
WERNER, D., and KISSLING, E. (1985), *Gravity Anomalies and Dynamics of the Swiss Alps*, Tectonophysics *117*, 97–108.

(Received April 29, 1998, revised November 17, 1998, accepted April 30, 1999)

 To access this journal online:
http://www.birkhauser.ch

Pure appl. geophys. 157 (2000) 131–146
0033–4553/00/020131–16 $ 1.50 + 0.20/0

| **Pure and Applied Geophysics** |

Upper Crustal Velocity Structure in Slovenia from Rayleigh Wave Dispersion

MLADEN ŽIVČIĆ,[1] ISTVÁN BONDÁR[2] and GIULIANO F. PANZA[3]

Abstract—The inversion of surface-wave dispersion curves can provide information on the average elastic properties of the upper crustal layers that are usually poorly sampled by body waves. The broad band digital records of earthquakes which recently occurred in Slovenia and neighbouring regions are used to extract the group velocity of the fundamental mode of Rayleigh waves, using frequency-time analysis (FTAN). The obtained dispersion curves permit a good resolution for the velocity and the thickness of the upper crust. The thickness of the uppermost sedimentary layer varies between 4 and 6 km and its shear-wave velocity is less than 3 km/s. The lower sedimentary layer is 7 to 9 km thick and its shear-wave velocity ranges from about 3.05 km/s in eastern Slovenia, to about 3.25 km/s in western Slovenia. The shear-wave velocity in the crystalline layer is around 3.5–3.7 km/s in the eastern part, while in the western part it reaches a rather high value of about 3.85 km/s.

Key words: Slovenia, upper crust, structure, model, surface waves.

Introduction

A seismogram contains information on the earthquake as an energy source, and information on the medium through which the radiated energy propagates. Due to the complex nature of the earthquake source and the complex structure of the earth as the propagation medium for elastic waves, it is usually very complex in its appearance. With a proper selection of the parts of an earthquake record and with a proper choice of the theoretical model at the base of the experiment design, it is possible to extract the relevant information on some parameters of earthquake source or earth structure. Thus extracting the fundamental mode of Rayleigh waves and determining its dispersion it is possible to estimate the average structure of the

---

[1] Geophysical Survey of Slovenia, Observatory, Pot na Golovec 25, SI-1000 Ljubljana, Slovenia, and Slovenian Association for Geodesy and Geophysics, Kersnikova 3/II, SI-1000 Ljubljana, Slovenia.

[2] Seismological Observatory, Hungarian Academy of Sciences, Meredek u. 18, H-1112 Budapest, Hungary, presently at Center for Monitoring Research, 1300 N 17th Street, Suite 1450, Arlington, VA 22209, U.S.A.

[3] Department of Earth Sciences, University of Trieste, Via E. Weiss 1, I-34127 Italy and the Abdus Salam International Center for Theoretical Physics, Trieste, Italy.

earth along the path that waves have travelled. The interpretation of phase velocity dispersion, as determined from records of a single station, requires the knowledge of the earthquake source mechanism, whereas group velocity dispersion curves are, within certain limitations, not influenced by the earthquake source mechanism.

Slovenia lies at the northeastern boundary of the Adria microplate (RAVNIK et al., 1995). Its territory comprises three major geotectonic units: the Alps, the Dinarides and the Pannonian basin. In the plate tectonic sense, it is situated at the collision margin of the Adriatic microplate with the Eurasian plate. For this reason it is structurally very complex. To the south the undeformed Istrian platform, together with the External Dinarides form the unique Adriatic-Dinaric carbonate platform that, to the north underthrusts the Southern Alps (Internal Dinarides). Further to the north, the Periadriatic line separates the Southern from the Eastern Alps. To the northeast these units are covered by the Tertiary molasse sediments of the Pannonian basin.

Velocity models of the earth's crust in Slovenia are practically non-existent. The research carried out to date is on a substantially larger regional scale and of low resolution: the territory of Slovenia has been included in the studies of crustal and upper mantle structure of large regions such as Europe (PANZA et al., 1980; SUHADOLC and PANZA, 1989), Central Europe and Mediterranean (SPAKMAN, 1990), Southeastern Europe (NESTEROV and YANOVSKAYA, 1988, 1991), Eastern Alps and Pannonian basin (BONDÁR et al., 1996), Mediterranean (CALCAGNILE and PANZA, 1990; PIROMALLO and MORELLI, 1997) and Balkans (ROTHÉ, 1972).

Only one deep seismic sounding (DSS) profile crosses Slovenia in the NNE–SSW direction and reaches the Mohorovičić discontinuity (JOKSOVIĆ and ANDRIĆ, 1983). On this profile only the uppermost several kilometers are distinguished from the rest of the crust, and the total crustal thickness has been determined to be between 42 km in the south, under the External Dinarides, to less than 30 km in the NE, under the Pannonian basin. Using the crustal thickness from this profile as a reference, Ribarič compiled a map of the crustal thickness using gravity and topographic data (RIBARIč, 1987). The deployment of quality broad band digital stations in Slovenia and in Friuli in NE Italy opens new possibilities to use records of regional medium size events to perform surface-wave dispersion studies for structural purposes.

For seismic hazard studies that use a deterministic approach based upon complete waveform modelling, it is essential that good structural models are available (PANZA et al., 1996). Especially important is the role of the upper crustal layers. The velocity models obtained by inversion of body wave data give rather poor results, because seismological stations are, as a rule, situated on rock sites with high velocity and density, and seismic body waves (both longitudinal and transversal) tend to sample only the faster layers, avoiding the slower layers in the upper crust. As a consequence, the velocity models resulting from body-wave studies are faster than the average structure and, when used as input in waveform modelling, usually cause a significant underestimate of the ground motion. Thus the main aim of the paper is determination of the velocity model to be used for waveform modelling.

Data and Methodology

In recent years several quality broadband digital instruments have been installed in the area. The records of the stronger local and regional earthquakes serve as a good data base for crustal structure studies from the inversion of surface-wave dispersion curves. Since surface waves are generated over a wide frequency spectrum, ranging over several decades, the best recording instruments are very broadband and broadband seismometers with high dynamic range digital acquisition equipment. For this study we have used the records of the digital broadband stations in Ljubljana (LJU) and Trieste (TTE). The instruments and station identifications are shown in Table 1.

To determine the upper crustal velocity structure in Slovenia we concentrate on event-station paths that mostly sample only one of the geotectonic units, and we consider only events that are well recorded at the broadband stations in the region (LJU, TTE). Paths crossing several tectonic regimes were excluded from the analysis.

To guarantee high signal-to-noise ratio only the stronger events are used. Since the epicentral distances considered are rather short, accurate determination of earthquake epicentres and origin times is of crucial importance. In this study the source-receiver distances range from 53 to 94 km and the epicentres are determined with a standard error of less than 5 km, which is on average 7% of the considered source-receiver distances. For group velocities ranging from 2.5 to 3.0 km/s, this location error corresponds to a velocity standard error of about 0.2 km/s, which is similar to the effect of the errors in the origin time. For these reasons we selected the longest possible source-receiver paths and used the best available body wave model (MICHELINI *et al.*, 1998) to define the locations and origin times of the analysed events. One of the most important factors in precisely delineating hypocentral position (and particularly its depth) is the distance to the nearest

Table 1

Station coordinates and the instruments used in the study

Station code	Latitude	Longitude	Seismometer	Data logger
LJU	46.0438	14.5274	WR-1	SSR-1
TTE	45.6597	13.7944	STS-1	Q680-V

Table 2

The parameters of the earthquakes, the recording station and the source-receiver distances used in the study

Path	Date	Origin time	Latitude [°N]	Longitude [°E]	M_{WA}	Station	Distance [km]
#1	27.04.1991	18:44:53.6	46.57	15.12	3.6	LJU	74
#2	05.10.1991	05:14:58.1	46.24	13.31	4.0	LJU	94
#3	11.03.1992	15:40:31.6	45.93	14.34	3.3	TTE	53
#4	29.05.1993	08:43:10.5	45.55	15.29	4.2	LJU	77

station. This fact additionally restricted the selection of the suitable events. The earthquakes finally selected, listed in Table 2, enabled us to represent the structure with four paths reported on Figure 1. In the same figure, the stations used in the hypocentre determinations as well as the boundaries of the main geotectonic units are shown.

After correcting for the instrumental response, vertical component records are processed using frequency-time analysis (LEVSHIN *et al.*, 1972, 1992). FTAN

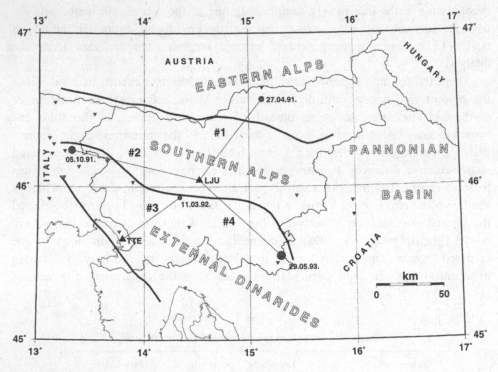

Figure 1

Epicentres of the four earthquakes used in the study (circles) and paths to the recording stations (large triangles). Major geotectonic units are indicated. Small inverted triangles denote the stations used in the epicentre locations.

represents a significant improvement, due to LEVSHIN *et al.* (1972, 1992), of the multiple filter analysis originally developed by DZIEWONSKI *et al.* (1969) and can be applied to a single channel to measure group velocity, and, if the source is known, phase velocity, even when there is higher modes contamination. FTAN employs a system of narrow-band Gaussian filters, with varying central frequency, that do not introduce phase distortion. For each filter band the square amplitude of the inverse FFT of the filtered signal is the energy carried by the central frequency component of the original signal. Since the arrival time is inversely proportional to group velocity, for a known distance, the energy is obtained as a function of group velocity at a certain central frequency. The process is repeated for different central frequencies. A floating filter, consisting in a sequence of frequency filters and time windows, is applied to the dispersion curve for an easy extraction of the fundamental mode. At this stage, the phase equalization and the inverse FFT operation give a pulse, which is isolated with a time window (RATNIKOVA, 1990) and the Rayleigh wave fundamental mode is extracted. The available data cover the period range from about 5 to about 20 s; in this period range and for the distances considered, the group velocity curves are practically independent from the source mechanism, while phase velocities can be severely distorted by the source apparent initial phase (PANZA *et al.*, 1973, 1975a,b); therefore we limit our analysis to group velocity measurements, determined together with a confidence range interval at each period. The period range covered by our dispersion measurements permits a limited depth penetration, and the depth range to which shear-wave velocity can be inverted is restricted to the upper crustal layers (PANZA, 1981). The depth of the Mohorovii discontinuity is taken from the map compiled by RIBARIČ (1987), and for the deeper layers a standard IASPEI91 model is used (KENNETT and ENGDAHL, 1991). The forward problem, i.e., the calculation of the theoretical dispersion curves, is solved using the Knopoff method (KNOPOFF, 1964) as described by SCHWAB and KNOPOFF (1972) and SCHWAB *et al.* (1984).

The inverse problem can be solved in several ways, mostly depending on the *a priori* knowledge of the model for which one attempts to invert. Here we have taken the approach that attempts to minimise the effect of the *a priori* knowledge of the structure. The nonlinear methods such as hedgehog, a variant of Monte Carlo search (VALYUS, 1972; PANZA, 1981), simulated annealing and genetic algorithms (DAVIS, 1990; GOLDBERG, 1989) provide a range of acceptable models. The first one does the selection within a given range of values, with discrete steps and tends to be time-consuming in the case of a large search space. The genetic algorithm approach starts from a random set of trial solutions and, by using the survival of the fittest principle as a selection criterion, generates a suite of models and iterates toward the optimum. Both the hedgehog and the genetic algorithm provide similar results. The hedgehog performs better in a smaller search space (inversions of up to 6–7 parameters) while the genetic algorithm may be computationally more efficient for a larger search space (BONDÁR *et al.*, 1996). Here we present results obtained by use of the genetic algorithms.

Table 3

The range of allowed values for the inverted parameters. The resolution is adjusted to satisfy the requirement that the number of steps is the power of 2

Layer	Depth to top of layer [km]	Resolution [km]	S-wave velocity [km/s]	Resolution [km/s]
1	0		1.0–4.0	0.024
2	0.0–10.0	0.32	2.0–4.0	0.016
3	10.0–20.0	1.43	3.0–4.5	0.024
4	20.0–40.0	1.33	3.5–5.0	0.021

The range of periods for which group velocity has been determined has the best resolution for intermediate crustal depths. Several attempts, made with different parameterisations, show that a four-layer's model, with variable thicknesses and velocities, corresponds best to the resolving power of the data, as we could expect from the synthetic tests made by PANZA (1981). To achieve a better resolution in the uppermost sedimentary layers we would need higher frequency data, while the longest observed period of 20 s does not allow the estimate of the total crustal thickness, which is kept fixed at 40 km, the average value given by RIBARIČ (1987).

To achieve a better constraint on the results, the individual measurement errors of group velocities at each period are built into the misfit function to be minimised during the inversion. The misfit of a solution is calculated as the weighted mean absolute difference of the observed and calculated fundamental mode Rayleigh wave group velocities, where the normalised weights are constructed from the measurement errors. In this way the misfit value not only characterises the fitness of the solution in general, but also forces the accepted solutions to be within the confidence limits at each measured period. The range of allowed values for the inverted parameters is given in Table 3.

Results

The seismological station in Ljubljana (LJU) has a central position within the country and the territory investigated. From the available braodband digital seismograms, we have selected the most representative (strongest) events with epicentre-station path in different parts of Slovenia. In Figures 2 to 5 we show the models that are consistent with the observed Rayleigh wave group velocities. The range shown corresponds approximately to twice the minimum misfit value. For path # 3 it has been possible to obtain a very low misfit (0.004) km/s), and we have plotted the models within 2.5 times the minimum misfit value. From these sets of allowed models we select the ones with median values of the parameters as the preferred ones and plot them as thick lines. In part b) of these figures the corresponding group velocity dispersion curves are plotted, the one corresponding to the preferred solution as a thick line.

An earthquake of $M_{WA} = 3.6$ in NE Slovenia (27.04.1991) was well recorded digitally in LJU (path #1). The path crosses the Periadriatic line and runs mostly through the Southern Alps (Fig. 1). The preferred model consists of an approximately 5-km thick layer of slow sediments with S-wave velocity around 2.75 km/s atop layers with S-wave velocity around 3.15 km/s. From the depth of about 16 km

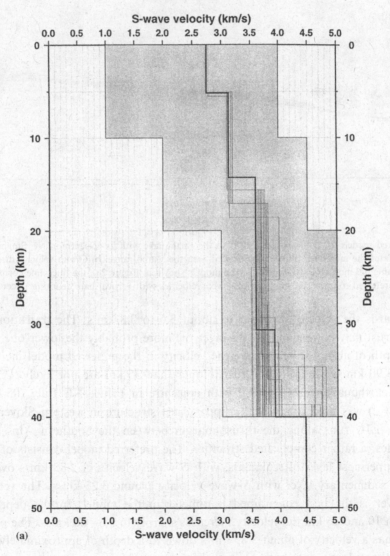

Fig. 2.

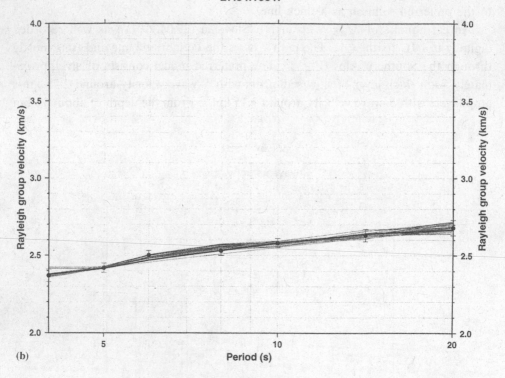

Figure 2

a) Accepted models for path #1 form LJU to the earthquake with its epicentre in NE Slovenia. Thick line indicates the preferred solution, the shaded area the search range. b) Group velocity dispersion of the fundamental mode Rayleigh wave for the models shown in Figure 2a), the thick line corresponds to the preferred solution and dots are observed velocities with vertical bars denoting uncertainty.

downwards the velocity increases to about 3.6 to 3.8 km/s. The transition to the lower crust may correspond to this layer or, more probably the lower one situated at a depth of about 31 km, where the velocity in the preferred model increases to about 3.90 km/s. Models with crust thinner than 35 km are not likely.

An earthquake of $M_{WA} = 4.0$ with epicentre in Friuli, NE Italy (05.10.1991) (path #2) was used to study the upper crust structure in western Slovenia. The path to LJU runs along the thrust margin between the Southern Alps and the Dinarides, a rather complicated structure. The preferred model consists of a 4-km thick uppermost layer of sediments, with S-wave velocity of 2.85 km/s overlaying a thick sedimentary layer with S-wave velocity around 3.25 km/s. The velocity of this layer is the best constrained parameter in the model. In the depth range between 10 and 16 km the velocity increases sharply to 3.6–4.0 km/s. The preferred model has a velocity of about 3.85 km/s down to a depth of approximately 31 km, where the velocity increases to mantle values. This increase, although poorly constrained, may correspond to the boundary between the upper European crust and the Adria mantle, as suggested for northeastern Italy by the results of deep seismic sounding (ITALIAN EXPLOSION SEISMOLOGY GROUP and INSTITUTE OF GEOPHYSICS, ETH, Zürich, 1981; SCARASCIA and CASSINIS, 1997).

Path #3 from an earthquake in central Slovenia (11.01.1992, $M_{WA} = 3.3$) to Trieste (TTE) lies entirely in the region of the External Dinarides. The preferred model consists of a 6-km thick surficial layer of sediments (V_s about 3.00 km/s)

05.10.1991.

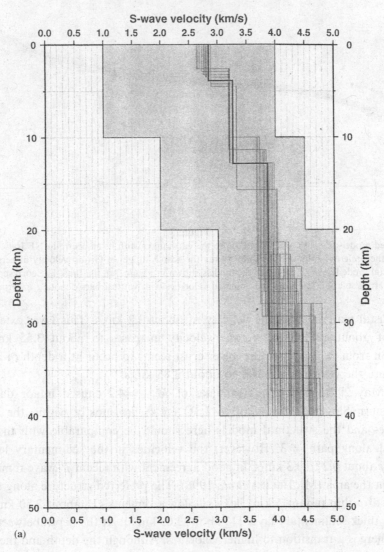

(a)

Fig. 3.

05.10.1991.

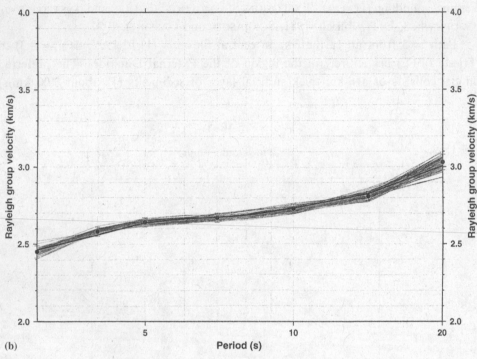

(b)

Figure 3

a) Accepted models for path #2 from LJU to the earthquake with its epicentre in NE Italy. Thick line indicates the preferred solution, the shaded area the search range. b) Group velocity dispersion of the fundamental mode Rayleigh wave for the models shown in Figure 3a), the thick line corresponds to the preferred solution and dots are observed velocities with vertical bars denoting uncertainty.

atop crustal material with the velocity of about 3.2 km/s. This layer extends to a depth of about 13 km, where the velocity increases to about 3.55 km/s. The transition from the upper to the lower crust seems to occur at a depth of about 27 km, where the velocity increases to about 3.85 km/s.

On May 29th, 1993 an earthquake of $M_{WA} = 4.2$ caused minor damage in southeastern Slovenia. The path to LJU (#4) lies mostly within the External Dinarides and the structural models here should be comparable with the models obtained along path #3. However, the velocities in the sedimentary layers are lower by about 0.2 to 0.3 km/s, in good agreement with recent P-wave tomographic studies in the area (MICHELINI *et al.*, 1998). The preferred structure along this path consists of a thin (about 5 km) layer of slow sediments (V_s about 2.80 km/s) atop a 10 km thick sedimentary layer (V_s about 3.05 km/s). At the depth between 10 and 15 km there is a transition to higher velocities. Although the depth and the velocity itself are not very well constrained, all models have a velocity jump of at least 0.3

km/s. This velocity contrast is similar to and occurs at similar depths as in the model for western Slovenia (path # 2).

11.03.1992.

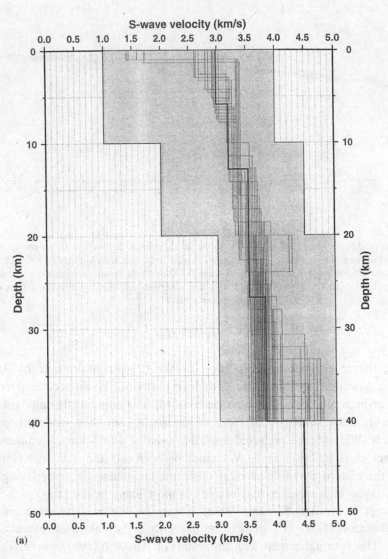

(a)

Fig. 4.

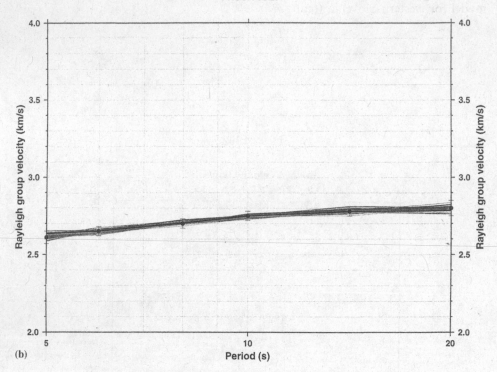

11.03.1992.

(b)

Period (s)

Figure 4

a) Accepted models for path #3 from TTE to the earthquake with its epicentre in central Slovenia. Thick line indicates the preferred solution, the shaded area the search range. b) Group velocity dispersion of the fundamental mode Rayleigh wave for the models shown in Figure 4a), the thick line corresponds to the preferred solution and dots are observed velocities with vertical bars denoting uncertainty.

Conclusions

Using the group velocity dispersion of the fundamental mode of Rayleigh waves, the S-wave velocity structure for four paths in Slovenia, corresponding to different geotectonic units, has been determined. The comparative analysis of the models reveals the similarity of three of the paths, only the westernmost being significantly different. All preferred models consist of 4 to 6 km of sediments with S-wave velocities below 3 km/s. Velocities between 3.0 and 3.3 km/s (somewhat lower in the eastern part of Slovenia) seem to characterise the underlaying 7 to 9 km thick layer. The upper crustal velocity is most likely in the range of 3.5 to 3.8 km/s, except in western Slovenia where it may be higher. The velocities in the Dinaric direction (NW–SE) seem to be lower than the velocities in the NE–SW direction. This is in agreement with the results of RIBARIČ (1987) who analysed *Sg* arrival times for earthquakes with focal depths between 10 and 15 km. The velocities and layer boundaries in the lower crust are less well determined. The preferred models indicate that the upper crustal thickness is of the order of 30 km, in agreement with the velocity model for the Balkan region (Dinarides) (ROTHÉ, 1972). The data do not allow us to determine the crustal thickness, except in the northeastern part, where a crust thinner than 35 km is quite unlikely.

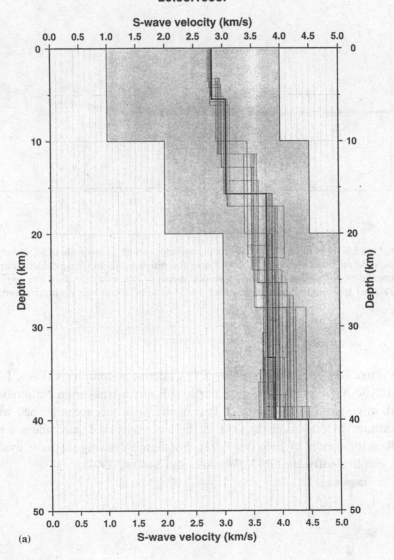

(a)

Fig. 5.

29.05.1993.

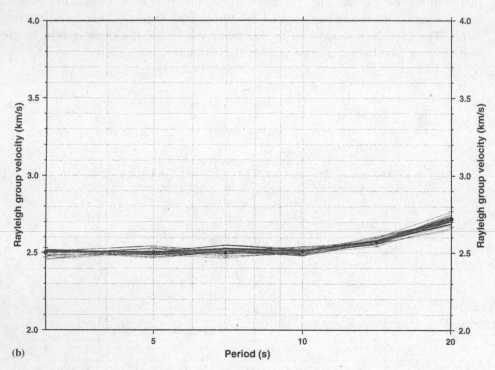

(b)

Figure 5

a) Accepted models for path #4 from LJU to the earthquake with its epicentre in SE Slovenia. Thick line indicates the preferred solution, the shaded area the search range. b) Group velocity dispersion of the fundamental mode Rayleigh wave for the models shown in Figure 5a), the thick line corresponds to the preferred solution and dots are observed velocities with vertical bars denoting uncertainty.

Acknowledgements

This work has been supported by EC Copernicus contract CIPA-CT94-0238, the UNESCO/IGCP project 414, and partly (I.B.) by a Hungarian National Science Foundation grant (OTKA-T014976). Part of the work was accomplished while one of the authors (M.Ž.) was at the DST at the University of Trieste under a Rector's grant (Reactor's letter of Feb. 26, 1993). For figure drawing we have used public domain graphics software GMT (WESSEL and SMITH, 1991).

Realistic Modelling of Seismic Input for Megacities and Large Urban Areas (project 414)

REFERENCES

BONDÁR, I., BUS Z., ŽIVČIĆ, M., COSTA, G., and LEVSHIN, A. (1996), *Rayleigh Wave Group and Phase Velocity Measurements in the Pannonian Basin*, Proceedings, XV Congress of the Carpatho-Balcan Geological Association, Athens, Sept. 17–20, 1995, 73–86.

CALCAGNILE, G., and PANZA, G. F. (1990), *Crustal and Upper Mantle Structure of the Mediterranean Area Derived from Surface-wave Data*, Phys. Earth Planet. Inter. *60*, 163–168.

DAVIS, L., ed., *Genetic Algorithms and Simulated Annealing, Research Notes in Artificial Intelligence* (Pitman, London 1990).

DZIEWONSKI, A., BLOCH, S., and LANDISMAN, M. (1969), *A Tectonic for the Analysis of Transient Signals*, Bull. Seismol. Soc. Am. *59*, 427–444.

GOLDBERG, D. E., *Genetic Algorithms in Search, Optimization and Machine Learning* (Addison-Wesley, Reading, MA 1989).

ITALIAN EXPLOSION SEISMOLOGY GROUP and INSTITUTE OF GEOPHYSICS, ETH, Zürich (1981), *Crust and Upper Mantle Structures in the Southern Alps from Deep Seismic Sounding Profiles (1977, 1978) and Surface-wave Dispersion Analysis*, Boll. Geofis. Teor. Appl. *92*, 297–330.

JOKSOVIĆ, P., and ANDRIĆ, B., *Ispitivanje građe zemljine kore metodom dubokog seizmičkog sondiranja na profilu Pula-Maribor* (Geofizika, Zagreb 1983) (in Croatian, unpublished).

KENNETT, B. L. N., and ENGDAHL, E. R. (1991), *Travel Times for Global Earthquake Location and Phase Identification*, Geophys. J. Int. *105*, 429–466.

KNOPOFF, L. (1964), *A Matrix Method for Elastic Wave Problems*, Bull. Seismol. Soc. Am. *54*, 431–438.

LEVSHIN, A. L., PISARENKO, V. F., and POGREBINSKY, G. A. (1972), *On a Frequency-time Analysis of Oscillations*, Ann. Géophys. *28* (2), 211–218.

LEVSHIN, A., RATNIKOVA, L., and BERGER, J. (1992), *Peculiarities of Surface-wave Propagation across Central Eurasia*, Bull. Seismol. Soc. Am. *82*, 2464–2493.

MICHELINI, A., ŽIVČIĆ, M., and SUHADOLC, P. (1998), *Simultaneous Inversion for Velocity Structure and Hypocenters in Slovenia*, Journal of Seismology *2*, 257–265.

NESTEROV, A. N., and YANOVSKYAYA, T. B. (1988), *Lateral Lithosphere Inhomogeneities in Southeastern Europe from Surface Wave Observations*, Izv. AN SSR, Fizika Zemli *11*, 3–15 (in Russian).

NESTEROV, A. N., and YANOVSKAYA, T. B. (1991), *Interferences on Lithospheric Structure in Southeastern Europe from Surface Wave Observations*, XXII General Assembly ESC, Proceedings and Activity Report 1988–1990, I, 93–98.

PANZA, G. F., *The resolving power of seismic surface waves with respect to crust and upper mantle structural models*. In *The Solution of the Inverse Problem in Geophysical Interpretation* (ed. Cassinis, R.) (Plenum Pub. Corp 1981) pp. 39–77.

PANZA, G. F., SCHWAB, F., and KNOPOFF, L. (1973), *Multimode Surface Waves for Selected Focal Mechanism. I. Dip-slip Sources on a Vertical Fault Plane*, Geophys. J. R. Astr. Soc. *34*, 265–278.

PANZA, G. F., SCHWAB, F., and KNOPOFF, L. (1975a), *Multimode Surface Waves for Selected Focal Mechanisms. II. Dip-slip Sources*, Geophys. J. R. Astr. Soc. *42*, 931–943.

PANZA, G. F., SCHWAB, F., and KNOPOFF, L. (1975b), *Multimode Surface Waves for Selected Focal Mechanisms. III. Strike-slip Sources*, Geophys. J. R. Astr. Soc. *42*, 945–955.

PANZA, G.F., MUELLER, ST., and CALCAGNILE, G. (1980), *The Gross Features of the Lithosphere-asthenosphere System in Europe from Seismic Surface Waves and Body Waves*, Pure appl. geophys. *118*, 1209–1213.

PANZA, G. F., VACCARI, F., COSTA, G., SUHADOLC, P., and FAEH, D. (1996), *Seismic Input Modelling for Zoning and Microzoning*, Earthquake Spectra *12*, 529–566.

PIROMALLO, C., and MORELLI, A. (1997), *Imaging the Mediterranean Upper Mantle by P-wave Travel Time Tomography*, Ann. Geofis. *XL* (4), 963–979.

RATNIKOVA, L. I., *Frequency-time analysis of surface waves, Workshop on Earthquake Sources and Regional Lithospheric Structures from Seismic Wave Data* (International Centre for Theoretical Physics, Trieste 1990) pp. 1–12.

RAVNIK, D., RAJVER, D., POLJAK, M., and ŽIVČIĆ, M. (1995), *Overview of the Geothermal Field of Slovenia in the Area between the Alps, the Dinarides and the Pannonian Basin*, Tectonophysics *250*, 135–149.

RIBARIČ, V. (1987), *On the Mohorovičić discontinuity in the Region of Slovenia*, Acta Geologica *17* (1–2), Zagreb, 21–30.

ROTHÉ, J. P., *Tables des temps de propagation des ondes séismiques (Hodochrones) pour la région des Balkans, Manuel d'utilisation* (BCIS, Strasbourg 1972).

SCARASCIA, S., and CASSINIS, R. (1997), *Crustal Structures in the Central-eastern Alpine Sector: A Revision of the Available DSS Data*, Tectonophysics *271*, 157–188.

SCHWAB, F. A., and KNOPOFF, L., *Fast surface wave and free mode computations*. In *Methods in Computational Physics, 11* (ed. Bolt, B. A.) (Academic Press, New York 1972) pp. 87–180.

SCHWAB, F., NAKANISHI, K., CUSCITO, M., PANZA, G. F., LIANG, G., and FREZ, J. (1984), *Surface-wave Computations and the Synthesis of Theoretical Seismograms at High Frequencies*, Bull. Seismol. Soc. Am. *74*, 1555–1578.

SPAKMAN, W. (1990), *Tomographic Images of the Upper Mantle below Central Europe and the Mediterranean*, Terra Nova *2*, 542–553.

SUHADOLC, P., and PANZA, G. F., *Physical properties of the lithosphere-asthenosphere system in Europe from geophysical data*. In *The Lithosphere in Italy, Advances in Earth Science Research* (eds. Boriani, A., Bonafede, M., Piccardo, G. B., and Vai, G. B.) (Acad. Naz. Lincei 1989) pp. 15–44.

VALYUS, V. P., *Determining seismic profiles from a set of observations*. In *Computational Seismology* (ed. Keilis-Borok, V. I.) (Consult. Bureau, New York 1972).

WESSEL, P., and SMITH, W. H. F. (1991), *Free Software Helps Map and Display Data*, EOS Trans. Amer. Geophys. U. *72*, 441, 445–446.

(Received May 5, 1998, revised November 27, 1998, accepted November 27, 1998)

To access this journal online:
http://www.birkhauser.ch

Pure appl. geophys. 157 (2000) 147–169
0033–4553/00/020147–23 $ 1.50 + 0.20/0

┃Pure and Applied Geophysics

Generalised Seismic Hazard Maps for the Pannonian Basin Using Probabilistic Methods

R. M. W. Musson[1]

Abstract—A set of seismic hazard maps, expressed as horizontal peak ground acceleration, is presented for a large area of Central Eastern Europe, covering the Pannonian Basin and surrounding area. These maps are based on (a) a compound earthquake catalogue for the region; (b) a seismic source model of 50 zones compiled on the basis of tectonic divisions and seismicity, and (c) a probabilistic methodology using stochastic (Monte Carlo) modelling. It is found that the highest hazard in the region derives from intermediate focus earthquakes occurring in the Vrancea seismic zone; here the hazard exceeds 0.4 g at return periods of 475 years. Special account has been taken of the directional nature of attenuation from this source. The maps are intended for use in studies of comparative methodologies for seismic hazard assessment.

Key words: Seismic hazard maps, Pannonian Basin, Monte Carlo simulation, Vrancea, intermediate focus earthquakes, seismic hazard methodology.

Introduction

The following report comprises part of the results of the Copernicus Project "Quantitative Seismic Zoning of the Circum Pannonian Basin" (EC Project CIPA-CT94-0238). In this part of the project, seismic hazard analyses of the study area are undertaken using two methodologies: a deterministic approach based on numerical synthesis of ground motion, and a probabilistic approach based on analysis of the regional earthquake catalogue through the medium of a seismotectonic seismic source zone model. This can be considered as a hybrid approach, in which the results for the two techniques are compared and contrasted. In this report, maps are produced using probabilistic methods.

The Earthquake Catalogue

The preparation of the earthquake catalogue and its properties have been documented in Musson (1996). A brief summary will be given here.

---

[1] British Geological Survey, West Mains Road, Edinburgh, EH9 3LA, UK.

In the absence of a homogenised catalogue for the whole region, based on original data treatment in a uniform way, it was necessary to compile a working catalogue from readily available sources. The regional limits of this catalogue are from 12–30 East and 42–50 North. The minimum magnitude is 4; three magnitude values are catered for in the dàta file (M_s, m_b and M_L) and an event accredited with a 4 magnitude in any of these three scales was included in the catalogue.

The parameters listed in the catalogue are as follows: day, month, year, hours, minutes, seconds, latitude, longitude, depth, M_s, m_b, M_L, agency code. The total number of earthquakes in the working file is 3946. The start date for the catalogue is nominally 1000; in fact the first earthquake occurs in 1022. The last event in the catalogue is dated 9 March 1993. From a statistical analysis, the catalogue as a whole is believed to be complete since 1885. For events larger than 5.5 M_s it is complete for another hundred years before that, and for the largest events (6.0 M_s) since 1590.

Since the publication of the catalogue minor revisions have been undertaken to correct errors that came to light subsequent to the report of MUSSON (1996). These mostly involved earthquakes on Romanian territory. Although other regional catalogues have been obtained (e.g., HERAK et al., 1996), a full repeat of the merging process would have delayed the project unacceptably. These new catalogues are themselves compilations, and there is a limit to the merits of producing compilations of compilations. More useful would be a regional catalogue produced entirely from a homogenous treatment of original data, in which parameters for all the earthquakes are produced in a consistent way. Such an initiative has been started in another project (STUCCHI, 1993).

Identification of Foreshocks and Aftershocks

Since probabilistic analysis of seismic hazard relies on the assumption that seismicity follows a Poisson process, it is generally considered essential to remove any non-Poissonian behaviour from earthquake catalogues. Any complete earthquake catalogue is clearly non-Poissonian: earthquakes are not entirely independent events because any substantial earthquake is usually followed by a cluster of aftershocks whose occurrence is dependent on the appearance of the main shock. If only main shocks are considered, then the earthquake behaviour for reasonably large areas is generally found to be described satisfactorily by the Poisson model (e.g., GARDNER and KNOPOFF, 1974; MARROW, 1992), in which case the use of hazard estimation models that assume a Poisson model is justified. The effect on the hazard estimation caused by the elimination of aftershocks from consideration is generally regarded as unimportant or acceptable on the grounds that aftershocks are an order of magnitude smaller than main shocks.

Since it is not valid to derive recurrence statistics from the complete catalogue and apply this to predicting main-shock occurrence, it is necessary to decluster the catalogue by removing all aftershocks and foreshocks (collectively referred to as accessory shocks). Otherwise, one will obtain incorrect estimates of the probability of large main shocks, since the slope of the magnitude-frequency curve will be affected by the appearance of many small events which are not main shocks (in effect, the removal of aftershocks makes the magnitude-frequency curve less steep).

To do this is not entirely a straightforward procedure. As is remarked by REASENBERG and JONES (1989), "aftershocks can only be identified in a statistical fashion: they bear no known characteristics differentiating themselves from other earthquakes." There are two ways in which this can be done: by manual inspection or algorithmically. To successfully remove accessory shocks by hand requires, ideally, first-hand knowledge of the earthquake catalogue on an event-by-event basis. This was practical, for example, in the hazard mapping study by MUSSON and WINTER (1997) for the UK, where one of the investigators was also the author of the earthquake catalogue used (MUSSON, 1994). In most cases, however, it is preferable to use some computational method for identifying accessory shocks, especially where earthquake catalogues are large.

The method used here is similar to that of REASENBERG (1985), although derived independently. The significant difference is that whereas Reasenberg's technique always considers the first event of a sequence to be the main shock, and a subsequent, larger earthquake becomes a "larger main shock" (REASENBERG and JONES, 1989), this method always considers the largest event in a sequence to be the main shock (if two equal events occur, the first is the main shock), enabling one to discriminate separately between foreshocks and aftershocks.

The algorithm works in the following manner: at the start, all earthquakes in the catalogue are flagged as "unassigned." The algorithm then considers each earthquake in descending order of magnitude. If it is unassigned, it must be a main shock (it has not been associated with a larger event). This earthquake is therefore flagged as "main shock," and the catalogue is then worked through backwards in time, looking at each event. If an event falls within the space/time window, it is flagged as a foreshock and the start of the time window reset. Once a period is found, equal in duration to the length of the time window, in which no foreshock can be identified, it is concluded that all foreshocks have been found. The same procedure is then used on looking for aftershocks, starting with the time of the main shock and working forwards in time through the catalogue. Through this sieving technique, all events in the catalogue are eventually identified as main shock, foreshock or aftershock. The computation is implemented by a computer program AFTERAN, which has options for either writing out a "pruned" catalogue of main shocks or conducting analyses of the aftershock behaviour in terms of the number of aftershocks observed for main shocks of differing magnitudes.

The size of the distance window used by the program was magnitude-dependent (see Table 1). A single value of 100 days was used for the time window. This value was arrived at by inspection of several clusters of seismicity in the catalogue in order to establish a useful value by direct observation. The declustered catalogue is shown in Figure 1.

The Seismic Source Model

The seismic source model produced for this study contains fifty source zones. The principles on which it is constructed are the subdivision of the study area into its component tectonic features, while reflecting the distribution of seismicity. The basis of the source zonation was the work accomplished by other members of the project, as described in MÂNDRESCU and RADULIAN (1996), SUHADOLC (1996), SUHADOLC and PANZA (1996), ŽIVČIĆ et al. (1996), MÂNDRESCU et al. (1997), ŽIVČIĆ and POLJAK (1997). In addition, notice was taken of the zonation of part of the area used for the Global Seismic Hazard Assessment Programme (GSHAP) as documented in GRÜNTHAL et al. (1996).

Some effort was required in order to arrive at a single, usable zone model for the whole region. While it was obviously desirable to keep as much as possible to the ideas expressed in the references above, some alterations were needed for the following reasons:

1. to reconcile interpretations where differences existed;
2. to ensure that the zonation adequately reflected the seismicity pattern;
3. to eliminate zones too small to be analysed;
4. to obtain a seamless coverage over the entire study area;

Table 1

Distance in km from main shock at which minor events are considered to be foreshocks or aftershocks, as a function of magnitude (after GARDNER and KNOPOFF, 1974)

Magnitude	Radius
4.0–4.4	30
4.5–4.9	35
5.0–5.4	40
5.5–5.9	47
6.0–6.4	54
6.5–6.9	61
7.0–7.4	70
7.5–7.9	81
8.0–8.4	94

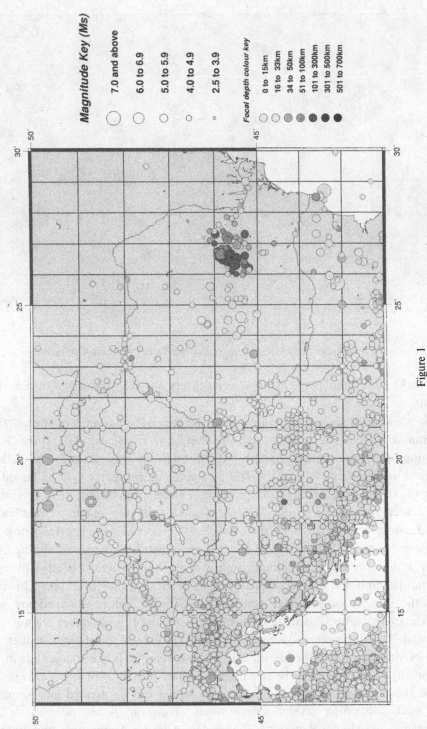

Figure 1

Seismicity of the study area. Earthquakes identified as foreshocks or aftershocks are not shown.

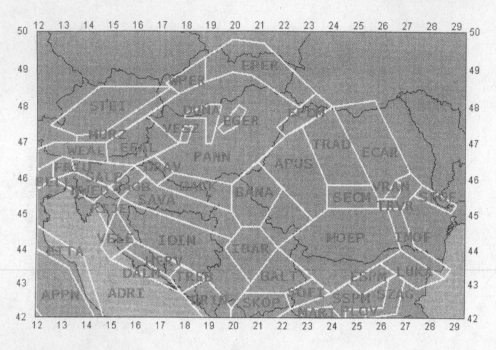

Figure 2
The seismic source zone model used in this study. See Table 2 for details of each zone.

5. to make such simplifications as were necessary for the production of a hazard map.

The zonation is shown in Figure 2, and the seismicity parameters of each zone are summarised in Table 2. Blanks in Table 2 indicate insufficient data. It should be born in mind when examining the depth distributions in Table 2 that some of them are based on very few earthquakes, and also some mean figures may be biased by values of 33 km which really indicate shallow indeterminate depths. The Gutenberg-Richter parameters a and b were calculated by the maximum likelihood method, using only portions of the catalogue considered complete following the analysis in MUSSON (1996). Parts of the frequency-distribution curve suffering from roll-off at low (or high) magnitudes were not used (they were excluded by eye). Uncertainties in a and b values were also calculated (but not listed here). In cases where the number of events is very low, clearly the values obtained are very approximate, however, in such low seismicity cases the precise values are not very influential. The maximum magnitudes shown in Table 2 are the observed values: the values eventually used in the hazard calculations were the largest magnitude in the zone or any zone deemed to be potentially similar, plus a small safety margin.

For the purposes of discussion, the zone model can be divided into six main areas, each containing several source zones. These areas are referred to, for convenience, as Alpine, Pannonian, Dinarides, Adriatic, Carpathian and Balkan.

Table 2

Summary statistics for the 50 seismic source zones used in the model

No.	CODE	Area	Start year	# Events	a	b	M_{max}	Events/ sq km	Mean depth	Max. depth
1	MURZ	10127	1267	50	3.223	−0.909	5.5	0.3815	12	43
2	WPER	6324	1443	13	1.288	−0.619	6.4	0.1026	12	36
3	EPER	29415	1453	20	3.144	−1.033	6.6	0.0349	10	40
4	STEI	27856	1590	17	3.575	−1.096	6.6	0.0557	12	20
5	WEAL	11552	1348	16	2.360	−0.900	5.7	0.0498	11	19
6	BELL	2863	1114	36	2.955	−0.906	6.8	0.7485	15	38
7	FRIU	3268	1278	65	2.717	−0.792	6.1	1.0832	13	35
8	SALP	9594	1077	76	4.249	−1.155	6.3	0.4436	11	35
9	PANN	90000	1092	21	2.706	−0.924	5.6	0.0114	12	19
10	VESZ	1428	1100	12	2.316	−0.900	5.6	0.3641	33	81
11	DUNA	5433	1258	17	2.958	−0.929	6.2	0.3213	11	17
12	EGER	3001	1767	9	2.933	−0.973	5.3	0.3662	9	16
13	ÈEAL	20015	1443	20	3.244	−1.022	6.5	0.0716	10	19
14	EPDM	8153	1781	19	2.830	−0.950	6.5	0.1314	16	50
15	DRAV	5653	1839	13	3.289	−1.015	5.6	0.2997	13	26
16	BACK	15197	1528	10	3.059	−0.969	5.7	0.1003	24	30
17	BANA	21501	1797	29	3.349	−0.966	5.6	0.1421	12	45
18	ZAGB	4862	1459	59	3.918	−1.045	6.2	1.1251	11	30
19	SAVA	27445	1502	34	3.434	−0.971	6.0	0.1293	13	33
20	NWED	7743	1279	37	2.172	−0.726	5.7	0.2394	12	30
21	RIJE	5641	1505	21	3.085	−0.906	5.8	0.5124	18	33
22	VELE	8620	1280	23	3.048	−0.973	5.7	0.1661	14	30
23	HERV	8259	1853	66	3.243	−0.848	6.1	0.8592	20	63
24	DALM	8300	1480	43	2.990	−0.848	6.1	0.4774	16	56
25	TREB	5662	1866	42	3.239	−0.906	6.1	0.7278	16	57
26	DRIN	12770	1563	63	3.597	−0.956	7.0	0.4643	15	46
27	IDIN	56817	1386	102	4.423	−1.091	6.0	0.2016	18	136
28	ADRI	83691	1123	114	3.738	−0.949	7.2	0.1045	24	72
29	EITA	17774	1268	99	4.036	−1.070	5.7	0.3208	17	48
30	APPN	27558	1243	277	3.888	−0.928	6.1	0.5442	14	59
31	APUS	32485	1614	6	2.735	−1.022	5.0	0.0137	20	33
32	TRAD	27889	1517	10	1.200	−0.680	5.6	0.0108		
33	ECAR	35449	1499	12	2.535	−0.876	5.0	0.0303	71	230
34	SECM	11059	1473	15	1.995	−0.777	6.4	0.0697	15	23
35	VRAN	1884	1022	143	4.375	−0.902	7.5	31.0398	127	165
36	TRVR	6897	1906	24	2.198	−0.670	.4	0.4779	64	150
37	MOEP	66997	1906	1			5.1			
38	IMOF	11369	1901	13	1.796	−0.634	7.2	0.1601	29	60
39	SFGF	4647	1901	8	2.930	−0.963	5.2	0.2575	24	46
40	SWCM	15599	1665	15	3.050	−0.981	5.6	0.0857	15	20
41	IBAR	17475	1739	84	3.264	−0.845	6.0	0.4381	14	43
42	BALT	17868	1902	10	1.824	−0.693	4.8	0.0631	11	22
43	SKOP	14091	1641	39	3.402	−0.927	5.8	0.3508	15	33
44	SOFI	7316	1818	20	2.464	−0.794	6.6	0.2653	26	77
45	SSPM	18935	1909	3	2.565	−0.950	4.9	0.0307	21	40
46	MARI	4000	1641	16	3.329	−0.975	6.9	0.6713	15	37
47	PLOV	2084	1750	18	2.506	−0.788	7.0	1.0842	17	49
48	ESPM	7502	1875	7	3.269	−0.958	7.0	0.3646	15	33
49	LUKA	12193	1892	8	2.216	−0.758	5.2	0.1253	16	51
50	SZAG	7970	1907	8	1.379	−0.550	5.9	0.1895	26	40

The division is not intended to be definitive; some zones are transitional and could be easily placed in more than one grouping.

Alpine Group

This area marks the northern boundaries of the Pannonian Fragment and the Adria Plate against the Eurasian Plate, summarised by GUTDEUTSCH and ARIC (1987). The Mur-Muerz Valley, running SW–NE through Austria, is an important locus of seismicity (zone MURZ). This lineation continues eastwards as the Peripieninian Lineament, which is here divided into two zones on the basis of seismicity (WPER and EPER).

To the south, the active centres of seismicity around Friuli and Belluno are well known and important to the hazard in this area. Each is given its own zone (FRIU and BELL). To the east of Friuli is the Southern Alps zone (SALP), a zone of moderate and shallow seismicity, the tectonics of which are described by ŽIVČIĆ and POLJAK (1997). Between these three zones and the Mur-Muerz Valley is inserted a zone of relatively low seismicity (WEAL).

The seismically active area in Austria north of the Mur-Muerz Valley is treated as a single zone (STEI). The seismicity here is lower than in the adjacent zones described above, although it has a larger magnitude earthquake within it (unless this event is mislocated). The area further to the north, on the Czech border, is of sufficiently low seismicity to be ignored altogether, especially as this area is marginal to this study.

Pannonian Group

This group of zones makes up the Pannonian Basin itself, which can be divided into a number of different features.

The largest zone (PANN) occupies the Great Hungarian Depression. The seismicity here is low to moderate, but clusters of higher activity can be discerned, especially in the area of Veszprem (VESZ), the bend of the Danube above Budapest (DUNA) and the Eger district (EGER). The westernmost extension of the Pannonian Fragment, between the Mur-Muerz Valley and the Peradriatic Lineament, includes the Mura Depression and is modelled as a single zone (EEAL).

The southern edge of this system mostly comprises basins with a WNW–ESE trend, becoming more NW–SE towards the west. These include the Drava, Backa Sava, Slavonia and Srem Depressions. The first two of these each are given one zone (DRAV, BACK). The remainder are similar in seismicity, and grouped together. To the east, the Banat Depression includes significant faulting with a roughly N–S trend (BANA).

A zone of seismicity with an apparent NE–SW trend is found to the northeast of the Pannonian Basin. This is the East Pannonian Basin Margin (EPBM).

Finally, the Zagreb-Balaton Zone (ZAGB) is characterised by a set of regional faults in a NE–SW direction with sinistral horizontal displacement (ROYDEN and HORVATH, 1988). This is a marginal zone, part Alpine, part Dinaric.

The Dinarides Group

The main division within this group is between the External and Internal Dinarides. The former set of zones (NWED, RIJE, VELE, HERV, DALM, TREB, DRIN) marks the collision zone of the Adria Plate on its NW margin. This is an area of considerable seismicity which, however, varies with local variations in the style of faulting, the predominant trend over the whole region being a dense pattern of NW–SE trending faults also with thrusting from NE to SW. The division into individual source zones is a simplified version of that proposed by MARKUSIC et al. (1996) as reported by SUHADOLC and PANZA (1996). The highest seismicity is found in the southernmost of these zones (DRIN), which marks an area of complexity on the flank of the Subpelagonic Massif, characterised by significant NE–SW trending faults intersecting with the characteristic Dinaric regional trend.

The Internal Dinarides represent a palaeo-subduction zone between the Adria Plate and the Pannonian Fragment (or microplate). As such, it includes deeper seismicity than is found elsewhere in the Dinarides. For the most part, this system, which preserves the Dinaric NW–SE trend, is modelled as a single zone (IDIN).

The Adriatic Group

The area of the Adriatic Sea and the Italian Peninsula is of marginal concern to the seismic hazard of the Pannonian area, and has therefore been treated in a very simplistic way. The Adriatic Sea has been treated as a single zone (ADRI), with Italy being divided into two zones: one occupying the eastern coast (EITA) and the remainder of this corner of the overall study area as a residual zone (APPN). Obviously, if one were concerned with sites in this area, a finer zonation would be required.

The Carpathian Group

The tectonics and seismicity of this area are discussed by MÂNDRESCU et al. (1997). The zones in this group cover the area around the Carpathian Mountains. In the north are three large zones of low seismicity: the Apuseni Mountains (APUS), the Transylvanian Depression (TRAD) and the Eastern Carpathians (ECAR). To the east of these lies the Moldavian Platform, which is of such low seismicity as to be almost aseismic, and is therefore excluded from the model. The same is true of the stable platform area in the Ukranian territory to the north. The Transylvanian Depression zone is unusual in that, although it contains a few historical earthquakes, the last recorded event in the catalogue is in 1902.

South of these three low-seismicity zones are three zones of considerably higher activity. These are the eastern extremity of the South Carpathians (SECM), the well-known Vrancea source, a subduction feature producing intermediate focus earthquakes (VRAN), and a transitional zone around the main Vrancea source to the east and south (TRVR) wherein the seismicity is more diffuse. The Vrancea earthquakes are of key importance to the seismic hazard of the region. As can be seen in Table 2, when seismicity rates are adjusted for area, the activity in the VRAN zone, square kilometre for square kilometre, is fifteen times greater than the next highest zone.

South again lies the Moesian Platform, a stable area of very low seismicity (MOEP). It would be possible to drop this zone from the model in the same way that the Moldavian Platform was omitted; however, it is not so peripheral, and thus is retained. The seismicity is too low to allow sensible analysis, therefore, semi-arbitrary values were assigned (these were actually imported from a neighbouring zone, SSPM, and reduced by not adjusting them for area).

The area from the Vrancean region to the Danube delta is more active than the territory to the north and south, and also coincides with the Sfântul Gheorghe Fault. This is modelled as a separate zone (SFGF), as is the course of the deep Intramoesian Fault which also appears to be an active feature (IMOF). The Intramoesian Fault continues southeastwards into the NE corner of Bulgaria (the Shabla region) where large earthquakes (7 M_s) have been observed. Modelling the entire fault as a single zone implies that such large earthquakes could also occur further north. This is conjectural and open to dispute; the significance of these two faults and the connection between the IMOF and the Shabla earthquake (if any) is poorly understood and was, for example, a subject of substantial discussion at the 1st International Workshop on Vrancea Earthquakes, Bucharest, 1–4 November, 1997. The area between the Sfântul Gheorghe Fault and the Intramoesian Fault is very inactive and has not been included in the model.

The Balkan Group

The zones of this group largely represent the Balkan Terrane, the western outer zone of the Protomoesian microcontinent, and parts of the Thracian Massif (HAYDOUTOV and YANEV, 1997). This area is characterised by a structural trend that continues N–S from the Banat Depression, and swings around the Moesian Platform, becoming NW–SE and finally almost E–W in central Bulgaria.

This group includes the western part of the South Carpathian Mountains (SWCM) on the NW flank of the Moesian Platform. To the SW of this zone is a complex area of relatively high seismicity, including part of the South European Variscan Suture (IBAR). To the east and southeast runs the Starayna Planina Meganticlinorum, south of the Moesian Platform and the Cis-Balcanic folded zone. The pattern of seismicity has been modelled as two relatively low-activity zones

(BALT, SSPM) and several local concentrations of higher activity (SKOP, SOFI, MARI, PLOV, ESPM and SZAG) of which three could be considered as continuing further south beyond the extent of the area covered by the earthquake catalogue. Finally, the Luda Kamčija Synclinorum appears to be a significant feature and occupies a single zone (LUKA).

Probabilistic and Stochastic Methodologies for Seismic Hazard

At the outset of this project it was intended that the software that would be used for the generation of seismic hazard maps would be the program SEISRISK III, a well-used and respected public domain program produced by the USGS (BENDER and PERKINS, 1987). However, as the project developed it became apparent that there were advantages in using a different approach.

The probabilistic method (often referred to as PSHA, standing for probabilistic seismic hazard assessment) follows these steps:

(i) The area to be studied is divided into discrete seismic source zones, each of which is deemed to be uniform in the character of its seismicity. There should be an equal probability that an earthquake of given magnitude could occur in any place within a single source zone. The geometry of the source zones is determined by the analysis from a combination of tectonic, geophysical, geological and seismological data.

(ii) The seismicity within each source zone is analysed, using the local earthquake catalogue, in order to determine the magnitude-frequency parameters and the maximum magnitude.

(iii) A locally appropriate attenuation relationship is chosen, to relate the expected ground motion at the site during an earthquake to the magnitude of the earthquake and its distance from the site. The uncertainty or scatter of the ground motion values regarding the predicted mean is an important variable which must also be used in the analysis.

(iv) The hazard analysis is based on the fact that the probability that an earthquake of magnitude m occurs in a source zone within any given distance interval is proportional to the fraction of the area of the zone that occurs within this range of the site. Since each source zone is deemed to be homogeneous, the fractional occurrences expected in any small sub-area of the zone can easily be calculated. An analytical integration is performed over all ground motion values, magnitudes, and source zones. From the results it is possible to determine the probability of any acceleration value being exceeded, assuming earthquake occurrences to follow a Poisson distribution.

The method is discussed in more detail elsewhere, for example REITER (1990).

A significant issue is how to treat uncertainties in the basic parameters of the seismicity distribution. The parameters of the magnitude frequency curve can be estimated from historical catalogues of earthquakes, although with a margin of error which may be significant. The maximum magnitude parameter is also uncertain, as no value can be proved to be a bounding limit. A means of dealing with uncertainties has been found following the work of COPPERSMITH and YOUNGS (1986). This is to use a logic tree in which different values for key parameters are assigned, with different weights attached to each. The problem inherent is that the choice of values and weights can be subjective and open to debate.

An alternative is to use a stochastic modelling approach also known as Monte Carlo simulation. In this approach, the initial three stages of the PSHA method are followed identically. The only difference is in the method by which the probabilities are calculated. Instead of analytical integration, the seismic source zone model is used to generate numerous synthetic earthquake catalogues, each having the same properties as the historical earthquake catalogue, but with random occurrences of earthquakes following the Poisson model, with epicentral locations randomly determined within each source zone. This process generates a very large number of synthetic observations at the site, and from these observations probabilities and return periods can be calculated directly in a simple and straightforward way.

This method has many advantages over conventional PSHA techniques. The first is the powerful way in which one can handle uncertainties in numerical parameters (such as activity rates or b values)—these can be modelled as distributions rather than discrete alternatives as in the logic tree approach. Each iteration of the program can sample a new value from the distribution. Secondly, special cases, such as non-Poissonian behaviour or local variations in attenuation can be dealt with very easily, whereas implementing such irregularities in an analytical PSHA can be difficult. Thirdly, on a non-technical note, the method can be very easily explained to, and understood by, those outside the seismological community such as politicians and planners who may have little mathematical ability. When it comes to the use of seismic hazard results for planning and policy making, it may be advantageous if those applying the results understand, and therefore have increased confidence in the methods used to obtain those results.

There is one disadvantage which possibly explains why the method is not more widely used, and that is that the most extreme values generated in the simulation process may vary from run to run, giving slightly different results. This is not really a significant problem as long as sufficient numbers of synthetic catalogues are used.

This technique is not new, and has been used in a number of seismic hazard studies in different parts of the world, as in studies by ROSENHAUER (1983), SHAPIRA (1983), JOHNSON and KOYANAGI (1988), AHORNER and ROSENHAUER (1993), etc.

In order to prove the applicability of the method and software (a program developed by BGS) to the present project, a validation exercise was undertaken. A section of the seismic source zone model was prepared for both SEISRISK III and the Monte Carlo Program in such a way that the input was to the extent possible identical. There are limits on the compatibility of the data files because of the different ways in which the programs operate, and some aspects of the source model had to be simplified to be used with SEISRISK (uncertainties were removed, depths were fixed at a constant 10 km). The zones used were STEI, MURZ, WPER, EPER, WEAL, EEAL, VESZ, DUNA, EGER, FRIU, SALP, ZAGB, DRAV, NWED, and RIJE. Two test sites were chosen: a high hazard site within zone SALP, and low hazard site not in any zone, but between zones EEAL and DUNA. (The zone PANN was not used because of the great difficulty of converting its complex geometry into the SEISRISK format.)

Comparing the output of different hazard programs and methods can be difficult, inasmuch as the output can be affected by the different ways in which the input data can be required to be entered, as has been shown by MAKROPOULOS *et al.* (1990). In this experiment it was necessary to convert the seismic source model into two quite differently structured input files to satisfy the two programs, disabling any feature in one program that cannot be adequately handled in the other (for example, SEISRISK is more or less confined to a single focal depth per seismic source zone, although there are possible workarounds for this).

One alteration to the SEISRISK code was necessary. The Monte Carlo hazard program used has a design limitation on the number of standard deviations (three) by which any calculated ground motion can scatter from the predicted mean. This is because very large scatter in the attenuation from the mean is believed to be physically unrealistic (e.g., REITER, 1990) and can have an undesirable effect on the hazard value at low probability levels. Exactly where to truncate the distribution is a matter of judgement; three sigma seems to be adequately conservative. SEISRISK normally allows an effectively unlimited scatter; this was altered to respect a three-sigma limit.

Hazard values were generated for two sites for peak ground acceleration values at three annual probability levels: 10^{-2}, 10^{-3} and 10^{-4}. The Monte Carlo software was instructed to produce synthetic catalogues of 100 years' length. The number of 100-year catalogues on which the hazard calculations were made was 10,000 (effectively 1,000,000 years of data), which is the value to be used in the actual hazard map production.

The results of the exercise are tabulated in Table 3.

In Table 3, the first row of results provides the answers obtained from SEISRISK III, while the second was obtained by the Monte Carlo process. These figures show that the two methods give almost identical results in most cases, the typical deviation being around 0.002 g. Larger variations can be obtained from SEISRISK by just varying the setup procedures, e.g., the intervals at which distances of magnitudes are discretised. The worst case, for no very obvious reason, is the lower seismicity site at 10^{-4} (only) where the disagreement is 0.009 g. From these results it can be concluded that the stochastic modelling approach is compat-

Table 3

Results for the comparison of methodologies for two test sites. Figures are peak ground accelerations in g

Method	Site 1 10^{-2}	Site 1 10^{-3}	Site 1 10^{-4}	Site 2 10^{-2}	Site 2 10^{-3}	Site 2 10^{-4}
SEISRISK	0.139	0.287	0.489	0.056	0.109	0.186
Monte Carlo	0.136	0.285	0.486	0.057	0.110	0.195

ible with conventional probabilistic methodology, and that the computer program used for this purpose is validated with respect to further stages of this project. However, in the rest of the project the model was used in full: the uncertainties in all parameters were incorporated. Each simulated catalogue sampled a value at random from the distributions of a and b values (independently), using the calculated uncertainties, and also for the M_{max} values (using an arbitrary standard deviation of 0.1 magnitude units), for each zone. Each simulated earthquake was given a depth drawn from the observed depth distribution for each zone. No correlation between depth and magnitude was assumed.

Attenuation

The remaining topic to be discussed is the attenuation relationship to be used. Since a well-known study for Europe (or more specifically, Southern Europe) exists in AMBRASEYS and BOMMER (1991), which includes strong motion data from Italy, Greece and the former Yugoslavia, it was considered to be adequate to the purposes of this study to adopt this formula, which is

$$\log_{10} A = -0.87 + 0.217 M_s - \log_{10} R - 0.00117 R \tag{1}$$

where A is peak horizontal ground acceleration in g, and R is hypocentral distance in km.

A more recent study also exists (AMBRASEYS, 1995) which draws upon an expanded database. The reasons for not using this study are (a) it does not operate on hypocentral distance, but rather fault rupture distance with a fixed depth factor; (b) there is little difference between the results of the two studies in practice.

While equation (1) is adequate for most of the study area, it reflects attenuation from crustal earthquakes and is not appropriate for intermediate-focus events from the Vrancea seismic zone. Furthermore, it is well known that the release of seismic energy from Vrancea events is markedly directional (e.g., MÂNDRESCU et al., 1988) in nature. Thus, the use of single isotropic attenuation equation such as that in equation (1) cannot provide realistic results.

The attenuation of strong ground motion from Vrancea intermediate-depth earthquakes is studied by LUNGU *et al.* (1997) from existing accelerogram data. Such data exist from three earthquakes only (1977, 1986, 1990) and for the first of these, only one data point (Bucharest) exists. LUNGU *et al.* (1997) present three directional analyses—for the direction of Moldova (NE), Chernavoda (SE) and Bucharest (S)—and one general case using all data irrespective of azimuth. The data point for Bucharest from the 1977 earthquake is added to the data sets for the Moldova and Chernavoda directions as otherwise there would be only two earthquakes and a regression would not be possible (Lungu, pers. com.).

The following equations are given:

$$\ln A = 5.571 + 0.937 M_s - 1.256 \ln R - 0.0069 h \tag{2}$$

$$\ln A = 6.470 + 0.923 M_s - 1.403 \ln R - 0.007 h \tag{3}$$

$$\ln A = 8.136 + 0.876 M_s - 1.675 \ln R - 0.0076 h \tag{4}$$

$$\ln A = 4.150 + 0.913 M_s - 0.962 \ln R - 0.006 h \tag{5}$$

where A is peak horizontal ground acceleration in cm/sec^2. These four equations relate to the complete data set, the Bucharest direction, the Cernavoda direction and the Moldova direction, respectively.

There are two problems with this suite of equations. In the first case, there is no term for anelastic attenuation. Where one would expect to find this, there is instead a term for depth. This means that differences in the azimuthal attenuation that relate to anelastic attenuation cannot be modelled properly. The second problem is that as the equations were all regressed separately, (and perhaps partly because there was no anelastic term in the model used) the equations are not compatible at short distances. It is expected that attenuation is slowest in the NE direction and much more rapid in the SE direction, and this is borne out by the coefficients for $\ln R$ (i.e., the geometric spreading term) in equations (4) and (5). However since the constant in equation (5) is considerably higher than that for equation (4), the net effect is that, for earthquakes at the critical depth of 90 km, for epicentral distances out to about 130 km, calculated ground motions are higher to the SE of the Vrancea zone than they are to the NE. The hazard map (not shown) that results from using these equations has, therefore, two ellipticalities: the contours of the highest hazard have their long axis towards the SE, while the contours of the lowest hazard have their long axis aligned to the NE. This is quite contrary to what is seen in isoseismal maps (MÂNDRESCU *et al.*, 1988), and would not be acceptable as a realistic map of hazard in the region.

There was not the possibility, within the scope of this project, to attempt to recalculate the regressions from the original data, which were not available. This could be done in the future, preferably with the addition of new data from further earthquakes, to explore the effects of using different functional forms, and to determine whether an anelastic attenuation term improves the results or not. For this project difficulty therefore arose: none of the published attenuation curves appeared to give reasonable results, and there were no data available against which to construct or test a new set of curves. In these difficult circumstances a stopgap solution was necessary, and after some experimentation the following course of action was adopted:

(i) Equations (2)–(5) were converted to give results in g rather than cm/sec^2.

(ii) The magnitude terms and geometric spreading terms were left untouched.

(iii) The depth term was removed from equations (3)–(5) and an anelastic attenuation term (on R) introduced in its place. This was set to an arbitrary small value for equations (3) and (4), and zero for equation (5).

(iv) The constant term in equations (3)–(5) was adjusted until each of them produced calculated accelerations at short epicentral distances that corresponded with those derived from equation (2).

In effect, the attenuation curve from the complete data set was used as an anchor for the other three, ensuring that the absolute values are sufficiently true to the original data set. The retention of the second two terms preserves the relative azimuthal variations in attenuation, and the addition of an arbitrary small anelastic term is a gesture towards what is probably the case, that anelastic attenuation is less in the north-easterly direction. The new coefficients are as follows:

$$\ln A = -1.15 + 0.923M_s - 1.403 \ln R - 0.0004R \tag{6}$$

$$\ln A = 0.33 + 0.876M_s - 1.657 \ln R - 0.0004R \tag{7}$$

$$\ln A = -3.1 + 0.913M_s - 0.962 \ln R \tag{8}$$

for the directions of Bucharest, Cernavoda and Moldova respectively, where A is now given in g.

It was assumed that attenuation to the NW is similar to that to the SE, i.e., equation (7) was used in this direction. This seems to be borne out by isoseismal maps of the major Vrancea events. For intermediary directions, the software used interpolated linearly between equations (6)–(8). It is to be emphasised that these equations are only intended as stopgaps, incorporating necessarily arbitrary decisions in order to obtain realistic results in the frame of the present project; future work should involve a detailed study of the original data to provide definitive equations.

Results

Hazard calculations were carried out for the area 14°–28°E, 43°–49°N. This is deliberately slightly smaller than the area covered by the catalogue, in order to leave a border zone free, in which the hazard might be affected by seismicity occurring outside the catalogue area. Results were calculated at a grid interval of 0.25 degrees of longitude (about 20 km) and 0.20 degrees of latitude (about 22 km).

The final hazard maps are shown in Figures 3–6.

The hazard maps show that several areas within the region have enhanced hazard. The first of these, obviously enough, is the Vrancea region, which generates the highest hazard values on each map. The highest contour values for return periods of 100, 475, 1000 and 3000 years are 0.25 g, 0.4 g, 0.5 g and 0.65 g, respectively. The shape of the contours for the Vrancea hazard are strongly influenced by the attenuation, extending strongly to the NE, but the simple shape provided by the attenuation variation is distorted to the S and W by the presence of adjoining source zones which contribute additional hazard.

The second highest hazard in the region, perhaps surprisingly, is found to the west of Zagreb. Unlike Vrancea, this is not an area commonly associated with large earthquakes. The largest historical earthquake in the ZAGB source zone is the 9 November 1880 earthquake with magnitude 6.2 M_s. However, there is a relatively large number of events occurring within this zone compared to its area. This can be seen in Table 2, where the area-adjusted seismicity rate is actually greater than the zone for Friuli. In addition to this, the seismicity is shallow, with over 60% of the earthquakes which have depth information being at less than 10 km focal depth.

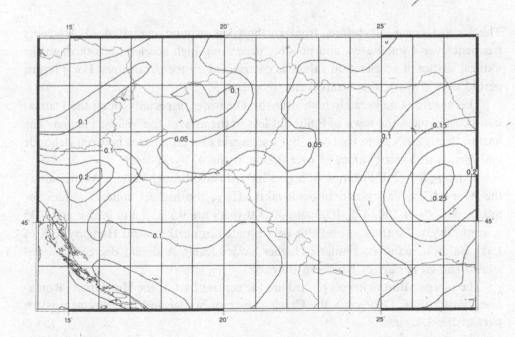

Figure 3
Contour map of horizontal peak ground acceleration values with return period of 100 years.

Figure 4
Contour map of horizontal peak ground acceleration values with return period of 475 years.

The hazard comes, therefore, from earthquakes around magnitude 5 M_s being frequent over a small area, and possibly generating high accelerations through the natural scatter of acceleration values as determined by the parameters. For a return period of 475 years, the hazard value reaches 0.35 g.

The hazard is moderately high over the Dinarides, especially along the Dalmatian coast around the town of Split, but less in the area of the Velebit Planina. The hazard drops away from the coast, but rises again in a broad area between Belgrade and Nis, with hazard values of over 0.25 g at the 475-year level.

A strong SW–NE lineation of higher hazard runs through eastern Austria along the Mur-Muerz Valley and into Slovakia. Here, the highest contour values for return periods of 100, 475, 1000 and 30,000 years are 0.1 g, 0.2 g, 0.25 g and 0.35 g, respectively. To the south of this belt, a patch in north-central Hungary, NE of Lake Balaton, manifests similar or higher hazard levels. A similar degree of hazard is also shown in part of Northern Bulgaria.

The places with the lowest hazard are the central part of the Hungarian–Romanian border (near Debrecen), the Czech–Austrian border and the extreme western part of the Ukraine.

The shape of the hazard contours is influenced by decisions made in the modelling process, and it is as well to be aware of where these may be subject to

Figure 5
Contour map of horizontal peak ground acceleration values with return period of 1000 years.

judgement, such that alternative interpretations might have a significant effect on the hazard. Such decisions must be made with respect to the purpose of the hazard study, and a generalised hazard mapping study, as here, requires a different approach from a site hazard mapping study as might be required to provide engineering parameters for a particular structure. This is discussed to good effect by PAGE and BASHAM (1985). In order to demonstrate hazard on a broad, regional scale, the assumption is generally made that large earthquakes such as the 14 June 1913 earthquake ($M = 7$) in northern Bulgaria are not restricted to the exact locations in which they occurred historically, but that other related features exist within the same geological structures which could produce similar earthquakes. The 1913 earthquake is therefore placed within the Eastern Starayna Planina Meganticlinorum zone, which is sufficiently broadly defined as to encompass some 7500 sq km. If this zone were drawn more tightly around the seismicity within this zone (which also includes event in 1875, 1914 and 1986) the effect would be to increase the hazard by concentrating it within a more constricted area. For a site-specific hazard study in this area, the effect of such an interpretation would have to be considered in the interests of conservatism. In a study such as the present one, which serves to present a regional outlook on hazard, there is not such a requirement to apply conservative options.

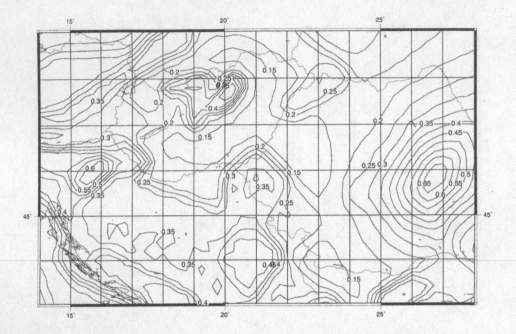

Figure 6
Contour map of horizontal peak ground values with return period of 3000 years.

There are a number of other areas where this applies; perhaps the most important is the Shabla region in NE Bulgaria, where the large 31 March, 1901 earthquake occurred. As discussed previously in this study, it is assumed that this earthquake is not fated to recur in one place only. An alternative interpretation would produce a peak of high hazard in the Shabla area. This possibility would need to be considered in the case of applications where the hazard in this area was particularly important. A less important case, but typical, is presented by Banja Luka, in which events of magnitude 5.3 occurred in 1969 and 1981, with smaller events taking place in 1897, 1935, 1950 and 1977. Modelling this activity as a small active zone would produce a peak of hazard at this location, the significance of which would be debatable.

Finally, considering that the hazard in the eastern part of the region is dominated by the Vrancea seismic source, there is still uncertainty about the directional attenuation from intermediate focus earthquakes. There are three points to be made in this respect. Firstly, as discussed above, the attenuation relations used in this study are partly artificially produced, and it is expected these could be improved with further work and more data. Secondly, little is known about attenuation to the W, SW and NW of Vrancea. Judging from isoseismal maps, it is likely that attenuation to the SW is less than simulated in this study, although more data is necessary before this can be quantified. Thirdly, some Vrancea earthquakes are known to have isoseismal elongation perpendicular to the usual axis, and also fault plane solutions different from the norm. These different events have not been factored into the hazard calculations. Probably they result from tearing within the subducting slab; they are therefore likely to be consistently smaller than other Vrancea earthquakes (limited rupture width) and so not taking account of them in the hazard calculations should not have too great an effect.

Conclusion

This study presents four maps of seismic hazard at different return periods for a wide area covering the Pannonian Basin and surrounds, including all of Hungary, Slovenia and Croatia, most of Romania and Serbia, and parts of Bulgaria, Slovakia, Austria and other adjoining countries. The hazard in the eastern part of the region is dominated by a single source of seismicity: the intermediate focus Vrancea activity. The western part is more complex, with the hazard being most acute along the Dalmatian coast and the Croatian–Slovenian border.

These maps are not intended for use in generating values for specific sites for engineering purposes, and are also not intended to supplant national maps of seismic hazard in any of the countries covered. National maps may implement specific policies for local purposes which are not considered here.

These maps can be used as a general guideline to regional distributions of seismic hazard in terms of peak ground acceleration values over the general area. They can also be used as a basis for comparative studies of seismic hazard in the area using different methodologies.

Acknowledgements

I would like to thank Paul Henni for his assistance in preparing the maps in this report, and to Dan Lungu, Gottfried Grünthal, Max Stucchi, and my various colleagues on the Copernicus Project "Quantitative Seismic Zoning of the Circum Pannonian Basin" (EC Project CIPA-CT94-0238) for helpful discussions at various stages of this work. This work was supported by the European Community and the Natural Environment Research Council and is published with the permission of the Director of the British Geological Survey (NERC).

REFERENCES

AHORNER, L., and ROSENHAUER, W., *Seismiche Risikoanalyse.* In *Naturkatastrophen und Katastrophen-vorbeugung,* Bericht zur Int. Decade Natural Disaster Reduction (Deutsche Forschungsgemeinschaft, Verlag, Weinheim 1993) pp. 177–190.

AMBRASEYS, N. N., and BOMMER, J. J. (1991), *The Attenuation of Ground Accelerations in Europe*, Eq. Eng. and Structural Dyn. *20*, 1179–1202.

AMBRASEYS, N. N. (1995), *The Prediction of Earthquake Peak Ground Acceleration in Europe*, Eq. Eng. and Structural Dyn. *24*, 467–490.

BENDER, B. K., and PERKINS, D. M. (1987), *SEISRISK III: A Computer Program for Seismic Hazard Estimation*, USGS Bulletin 1772.

COPPERSMITH, K. J., and YOUNGS, R. R., *Capturing uncertainty in probabilistic seismic hazard assessments within intraplate tectonic environments*. In *Proc. Third US Nat. Conf. Eq. Eng.* (Charleston 1986) *1*, 301–312.

GARDNER, J. K., and KNOPOFF, L. (1974), *Is the Sequence of Earthquakes in Southern California, with Aftershocks Removed, Poissonian?*, Bull. Seismol. Soc. Am. *64*, 1363–1367.

GRÜNTHAL, G., BOSSE, C., MUSSON, R. M. W., GARIEL, J.-C., DE CROOK, T., VERBEIREN, R., CAMELBEECK, T., MAYER-ROSA, D., and LENHARDT, W., *Joint seismic hazard assessment for the central and western part of GSHAP Region 3 (Central and Northwest Europe)*. In *Seismology in Europe* (ed. Thorkelsson, B.) (Icelandic Met. Office, Reykjavik 1996) pp. 339–342.

GUTDEUTSCH, R., and ARIC, K., *Tectonic block models based on the seismicity in the East Alpine-Carpathian and Pannonian area*. In *Geodynamics of the Eastern Alps* (eds. Flügel, H. W., and Faupl, P.) (Deuticke, Vienna 1987).

HAYDOUTOV, I., and YANEV, S. (1997), *The Protomoesian Microcontinent of the Balkan Peninsula—A Peri-Gondwanaland Piece*, Tectonophysics *272*, 303–313.

HERAK, M., HERAK, D., and MARKUSIC, S. (1996), *Revision of the Earthquake Catalogue and Seismicity of Croatia 1908–1992*, Terra Nova *8*, 86–94.

JOHNSON, C. E., and KOYANAGI, R. Y (1988), *A Monte-Carlo Approach Applied to the Estimation of Seismic Hazard for the State of Hawaii*, Seism. Res. Letters *59* (1), 18.

LUNGU, D., CORNEA, T., ALDEA, A., and ZAICENCO, A., *Basic representation of seismic action*. In *Design of Structures in Seismic Zones* (eds. Lungu, D., Mazzolani, F., and Savidis, S.) (Bridgeman Ltd, Timisoara 1997) pp. 9–60.

MAKROPOULOS, K., VOULGARIS, N., and LIKIARDOPOULOS, N., *A multi-methodological approach to seismic hazard assessment: an application for Athens (Greece)*. In *Proc. XXII General Assembly of the ESC* (Barcelona 1990) *2*, 585–590.

MÂNDRESCU, N., ANGHEL, M., and SMALBERGHER, V., *The Vrancea intermediate-depth earthquakes and the peculiarities of the seismic intensity distribution over the Romanian territory*. In *Recent Seismological Investigations in Europe* (ed. Neresov, I. L. et al.), *Proc XIX ESC General Assembly* (Moscow 1988) pp. 59–65.

MÂNDRESCU, N., and RADULIAN, M. (1996), *Characterisation of Seismogenic Zones of Romania*, EEC Technical Report, Project CIPA-CT94-0238.

MÂNDRESCU, N., POPESCU E., RADULIAN, M., UTALE, A., and PANZA, G. (1997), *Seismicity and Stress Field Characteristics for the Seismogenic Zones of Romania*, EEC Technical Report Project CIPA-CT94-0238.

MARKUSIC, S., COSTA, G., VACCARI, F., SUHADOLC, P., and HERAK, M. (1996), *Deterministic Seismic Zoning of the Croatian Territory Derived from Complete Synthetic Seismograms*, in preparation (referred to by Suhadolc and Panza 1996, vide infra).

MARROW, P. C. (1992), *Average Earthquake Recurrence Statistics for the UK Area and the Poisson Assumption*, BGS Technical Report No. WL/92/53.

MUSSON, R. M. W. (1994), *A Catalogue of British Earthquakes*, BGS Technical Report No. WL/94/04.

MUSSON, R. M. W., *An earthquake catalogue for the Circum-Pannonian Basin*. In *Seismicity of the Carpatho-Balcan Region, Proc. XV Congress of the Carpatho-Balcan Geol. Ass.* (eds. Papanikolaou, D., and Papoulia, J.) (Athens 1996) pp. 233–238.

MUSSON, R. M. W., and WINTER, P. W. (1997), *Seismic Hazard Maps for the UK*, Natural Hazards *14*, 141–154.

PAGE, R. A., and BASHAM, P. W. (1985), *Earthquake Hazards in the Offshore Environment*, U.S. Geological Survey Bulletin no. 1630.

REASENBERG, P. (1985), *Second-order Moment of Central Californian Seismicity, 1969–1982*, J. Geophys. Res. *90*, 5479–5495.

REASENBERG, P., and JONES, L. M. (1989), *Earthquake Hazard after a Mainshock in California*, Science *243*, 1173–1176.

REITER, L., *Earthquake Hazard Analysis* (Columbia UP, New York 1990).

ROSENHAUER, W., *Methodological aspects encountered in the Lower Rhine area seismic hazard analysis*. In *Seismicity and Seismic Risk in the Offshore North Sea Area* (eds. Ritsema, A. R., and Gürpinar) (Reidel, Dordrecht 1983) pp. 385–396.

ROYDEN. L.H. and HORVÁTH. F. (editors) (1988), *The Pannonian Basin*, Am. Assoc. Pet. Geol., Mem. *45*, 27–48.

SHAPIRA, A. (1983), *Potential Earthquake Risk Estimations by Application of a Simulation Process*, Tectonophysics *95*, 75–89.

SUHADOLC, P. (1996), *Structural Models in the Eastern Alps and Bulgaria*, EEC Technical Report, Project CIPA-CT94-0238.

SUHADOLC, P., and PANZA, G. (1996), *Focal Mechanisms and Seismogenetic Zones*, EEC Technical Report, Project CIPA-CT94-0238.

STUCCHI, M. (1993), *Recommendations for the Compilation of a European Earthquake Catalogue, with Special Reference to Historical Data prior to 1900*, EC Project RHISE "Review of Historical Seismicity in Europe", unpublished draft.

ŽIVČIĆ, M., CECIĆ, L., and POLJAK, M. (1996), *Seismicity and Geodynamics of Slovenia*, EEC Technical Report, Project CIPA-CT94-0238.

ŽIVČIĆ, M., and POLJAK, M. (1997), *Seismogenetic Areas of Slovenia*, EEC Technical Report, Project CIPA-CT94-0238.

(Received April 25, 1998, revised July 1, 1998, accepted July 1, 1998)

To access this journal online:
http://www.birkhauser.ch

Pure appl. geophys. 157 (2000) 171–184
0033–4553/00/020171–14 $ 1.50 + 0.20/0

| Pure and Applied Geophysics |

Seismic Zoning of Slovenia Based on Deterministic Hazard Computations

MLADEN ŽIVČIĆ,[1] PETER SUHADOLC[2] and FRANCO VACCARI[3]

Abstract—Seismic hazard of the territory of Slovenia is estimated using a deterministic approach based on the computation of complete synthetic seismograms. The input data are the catalogues of earthquakes and fault plane solutions for Slovenia and surrounding regions. Structural models are defined based on available seismological and geophysical information, but are mainly constrained by surface-wave dispersion and 3-D tomographic modelling of the upper crust. Seismogenic zones are delineated considering geotectonic characteristics, fault plane solutions and distribution of earthquake hypocentres. Outside Slovenia seismogenic zones are extended up to distances from which they can considerably influence seismic hazard estimates.

Synthetic seismograms are computed using the "receiver" structure along the entire path by normal mode summation (up to 1 Hz) for receiver sites on a 0.2 × 0.2 degrees grid and scaled to the magnitude of the earthquake allowing for spectral falloff. At each site the maximum value of horizontal velocity, horizontal displacement and design ground acceleration are considered as hazard parameter. The highest values are obtained for western Slovenia where the hazard is controlled by the strongest earthquake in the catalogue, the "Idrija" event of March 26, 1511.

Key words: Seismic hazard, seismic zoning, Slovenia.

Introduction

Studies of the seismic hazard of Slovenia were mostly performed for regulatory purposes. The first seismic map of Slovenia was proposed in 1963 (BUBNOV, 1996) and was based on maximum observed intensities. This map was further improved within the UNESCO/UNDP project on seismicity of the Balkan region (CVI-JANOVIĆ, ed., 1974). RIBARIČ (1986) published a revised map of maximal observed intensities in Slovenia and it can be considered the last published seismic hazard map based on a deterministic approach. A probabilistic approach was used for the first time when compiling the 1981 Seismic Building Code, an integral part of which was also a seismic hazard map for the territory of former Yugoslavia. For

[1] Geophysical Survey of Slovenia, Observatory, Pot na Golovec 25, SI-1000 Ljubljana, Slovenia and Slovenian Association for Geodesy and Geophysics, Kersnikova 3/II, SI-1000 Ljubljana, Slovenia.
[2] Department of Earth Sciences, University of Trieste, Via E. Weiss 1, I-34127 Trieste, Italy.
[3] Department of Earth Sciences, University of Trieste, Via E. Weiss 1, I-34127 Trieste, Italy and CNR-GNDT, Gruppo Nazionale per la Difesa dai Terremoti, via Nizza 128, I-00198 Roma, Italy.

Slovenia the Gumbel distribution of extremes was used to estimate expected intensities for return periods of 50, 100, 200, 500, 1000 and 10,000 years (ZAJEDNICA ZA SEIZMOLOGIJU SFRJ, 1987) LAPAJNE *et al.* (1995) published the first map consistent with the European prestandard Eurocode 8, using the probabilistic approach of CORNELL (1968). Later on, LAPAJNE *et al.* (1997a,b) proposed several modified versions based on recent developments of seismic hazard assessment (FRANKEL, 1995).

The seismic hazard of Slovenia was also considered in regional studies. SLEJKO and KIJKO (1991), in their study of seismic hazard of the main seismogenic zones in the Eastern Alps, included a large part of Slovenia in Ljubljana and Rijeka seismic zones.

The main objective of this study is to assess the appropriateness of the deterministic approach of COSTA *et al.* (1993), considering the size of the territory and the level of seismicity, as well as the level of detail with which seismic hazard can reasonably be mapped using this methodology.

Seismicity

Slovenia lies at the northeastern rim of the Adriatic microplate. Its seismicity is controlled by the geodynamic setting of the country within three large geotectonic units: the Alps, the Dinarides and the Pannonian basin, and is mainly constrained to the regions where these units are in direct contact.

Studies of seismicity of Slovenia (as well as that of neighbouring countries) rely mostly on macroseismic data (RIBARIČ, 1982). Until the late 1980s the number and distribution of seismological stations in Slovenia did not allow for the reliable estimate of earthquake parameters from instrumental records. It is estimated that the catalogue can be considered reasonably complete for magnitudes above 3.7 for the period after 1870 (LAPAJNE *et al.*, 1997b). As the largest magnitude since 1870 is estimated (using macroseismic data) to be 6.1 (the Ljubljana 1895 event), the range of completeness and the territory covered do not allow a reliable use of statistical methods (e.g., KRINITZSKY, 1995).

A good earthquake catalogue is required for every seismic hazard assessment. Many parts of Slovenia suffered major damage from earthquakes that originated outside its political boundaries. The strongest are the 1976 Friuli earthquake and two historical events in the Friuli-Carnia region, in 1348 and 1690. For the purpose of the seismic hazard assessment for the territory of Slovenia, we assembled a catalogue covering the region between 44.5–47.5°N and 12.5–17.5°E. The existing catalogues for the regions that surround Slovenia are rather inhomogeneous as regards their completeness as well as the methods used to derive earthquake parameters (CVIJANOVIĆ, 1981; ZSÍROS *et al.*, 1988; FIEGWEIL, 1981; OGS, 1992; and references therein). Therefore, the individual catalogues have been merged following the principle to retain the parameters determined in each catalogue only for events lying on the respective national territory. We therefore use ZSÍROS *et al.* (1988) parameters for events originating in Hungary and so on. Intensity is used as a measure of earthquake size for pre-1901 earthquakes. MCS and MSK scales used in different sources are assumed to be equivalent (WILLMORE, ed., 1979). For post-1900 earthquakes magnitude M_{LH} as defined by KARNÍK (1968) is used since the majority of earthquake magnitudes was determined uniformly when compiling the KARNÍK (1968, 1971) catalogue and the catalogue of earthquakes in Balkan Region (SHEBALIN *et al.*, 1974). Our merged catalogue contains earthquakes in the region 44.5–47.5°N and 12.5–17.5°E and we estimate it to be complete for intensities VI MSK and greater since 1890. However for deterministic hazard estimation purposes, only locations and magnitudes of the strongest earthquakes are of importance.

In estimating earthquake hazard it is of great importance to have a unified measure of the earthquake size. For historical events no instrumental data are available and magnitudes have to be estimated from macroseismic data. The relations between magnitude M_{LH} and isoseismal radii as determined from good quality isoseismal maps, have been derived from 20th century earthquake data.

For eighteen earthquakes both isoseismal maps and M_{LH} magnitudes are available. This has enabled us to derive relations for the estimation of magnitudes from macroseismic data. Isoseismals have been digitised and the relations between magnitude M_{LH} and equivalent radii R of isoseismals, as well as between M_{LH} and epicentral intensity Io and focal depth h, are derived:

$$M_{LH} = 2.72 + 1.63 \log R_{\mathrm{VI}} \quad \text{(correlation coefficient } r = 0.82\text{),}$$

$$M_{LH} = 1.08 + 2.32 \log R_{\mathrm{V}} \quad (r = 0.82),$$

$$M_{LH} = 1.14 + 1.96 \log R_{\mathrm{IV}} \quad (r = 0.84) \text{ and}$$

$$M_{LH} = 0.09 + 0.494 \, Io + 1.27 * \log(h) \quad (r = 0.74).$$

The relations are derived for intensities IV, V and VI MSK because there are not enough strong earthquakes in the 20th century which would provide data for calibrating magnitude formulas for higher intensities. The strongest earthquake in the 20th century was of magnitude $M_{LH} = 5.7$ and intensity VIII MSK. Magnitudes of historical events are therefore estimated from macroseismic data. The average value obtained using the above relations is adopted as macroseismic magnitude M_M. For that purpose we considered all earthquakes for which macroseismic data as well as instrumentally determined magnitude are available. The macroseismic magnitudes, M_M, of the two strongest earthquakes are:

$$26.03.1511. \quad M_M = 6.8 \pm 0.3$$

$$14.04.1895 \quad M_M = 6.1 \pm 0.2.$$

The 1976 earthquake with epicentre in Friuli (Italy) was used for control—its macroseismically estimated magnitude is $M_M = 6.1 \pm 0.1$, whereas its instrumental magnitude estimates range from $M_b = 5.9$ to $M_s = 6.5$ (ISC, 1976).

Seismogenic Structures and Earthquake Mechanisms

The data from the revised earthquake catalogue, as well as both published and newly determined fault plane solutions, were combined with the available geological and geophysical information to delineate and characterise seismogenic areas of Slovenia (POLJAK *et al.*, 2000). The main criteria were the geotectonic characteristics, fault plane solutions and earthquake hypocentres distribution. The studied region was divided into seven seismogenic zones (Fig. 1) having different tectonic

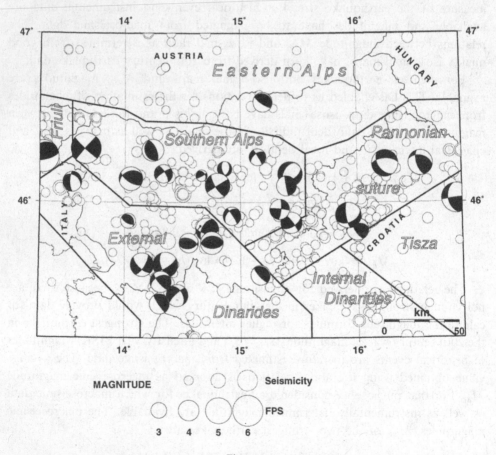

Figure 1
Seismogenic zones in Slovenia and fault plane solutions of earthquakes. Seismicity of $M > 3.7$ in the 567–1995 period is plotted as background. Note that magnitude scales are different.

evolution and seismicity characteristics. The zoning corresponds to a large extent to the structural geometry of the major geotectonic units. The entire investigated region is dominated by strike slip (NW–SE and NE–SW oriented) and thrust-type fault plane solutions, although normal type faulting is also present. For each seismogenic zone a characteristic earthquake mechanism is adopted. As a rule it corresponds to the strongest event for which the fault plane solution is available. For the Eastern Alps and the Internal Dinarides area the earthquake mechanism is assigned based on tectonic data.

Most of the seismicity in the area is constrained to be located in the upper crust, focal depths of recent events exceeding 15 km occur only in restricted areas. Due to the lack of reliable information on focal depths of historical events, the focal depth of 10 km was adopted for all events.

Structural Models

There are practically no velocity models pertinent to Slovenian territory. The research done so far, on a much larger regional scale and with low resolution, describes the structure of large regions like Europe (PANZA et al., 1980; SUHADOLC and PANZA, 1989), Southeastern Europe (NESTEROV and YANOVSKAYA, 1988, 1991), Alps, Pannonian basin (BONDÁR et al., 1996), Mediterranean (CALCAGNILE and PANZA, 1990) and Balkans. Only one deep seismic sounding profile crosses Slovenian territory (JOKSOVIĆ and ANDRIĆ, 1983), however its data resolve only the uppermost about 3 to 6 kilometres from the rest of the crust. Also, the total crustal thickness is determined to be between 42 km in the south under the External Dinarides and less then 30 km in the northeast.

Good structural models are essential for seismic hazard computations based on a deterministic approach using complete synthetic seismograms (PANZA et al., 1996). The role of the upper crustal layers is especially important. For that purpose velocity models obtained from body wave data give rather poor results—seismological stations are as a rule situated on rock sites with high velocities and density, and seismic waves (both longitudinal and transversal) sample only faster layers. Slower structures in the upper crust are only poorly sampled or not sampled at all. As a consequence the velocity models resulting from body wave data inversions are faster than the average structure and in waveform modelling usually result in unrealistically small amplitudes.

The structural regionalization of Slovenia is based primarily on surface wave data (BONDÁR et al., 1996; ŽIVČIĆ et al., this issue; COSTA et al., 1993) and on a tomographic inversion of longitudinal waves (MICHELINI et al., 1998). Both these models describe only layers in the upper crust. The depth of the crust is taken from the maps of the depth of the Mohorovičić discontinuity (RAVNIK et al., 1995 and references therein) while the upper mantle structure is assumed to be the same as in

AVERAGE STRUCTURAL MODELS

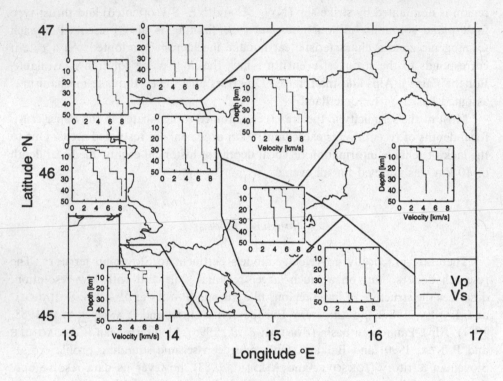

Figure 2
Structural regions used for synthetic seismogram computation with *P*- and *S*-velocity distribution in the crust.

the standard IASPEI91 model (KENNETT and ENGDAHL, 1991). Based on these considerations, the study area is divided into eight units which have different seismic velocity structural models (Fig. 2).

Method

The deterministic seismic hazard assessment consists basically of three steps. First, pertinent structures are identified and the location and the parameters of potential earthquakes are established. Then, a relation estimating ground motion parameters from earthquake magnitude (or epicentral intensity), epicentral distance and site conditions is determined/selected. Finally, parameters representing seismic hazard are selected and their maximum value for each grid point computed. Probabilistic approach differs in the third step when the probabilities of the exceedance are computed rather than maximum values. In our deterministic procedure, fully described in COSTA et al. (1993), the second step is replaced by numerical calculation of complete synthetic seismograms given the source parameters, distance and earth structural model. In this paper we have selected the maximum historical earthquake observed in a seismogenic zone as the maximum possible event for that zone.

The territory under investigation is divided into eight polygons and to each of them a uniform velocity structural model is assigned. At this stage only a bedrock type of structure is considered and surface layers with S-wave velocities lower then 1 km/s are not included. The same territory is divided in terms of the seismotectonic characteristics into seven seismogenic zones, and to each of them a characteristic focal mechanism is assigned. The delineation of seismogenic zones is based on the geodynamic model proposed by POLJAK et al. (2000), fault plane solutions and seismicity distribution. Outside Slovenia seismogenic zones were extended only to distances from which they can considerably influence seismic hazard estimates on the territory of Slovenia.

The catalogue of earthquakes for the period 1000–1995 is discretized over a $0.2° \times 0.2°$ grid retaining the largest magnitudes within each cell. To allow for location uncertainties of historical events, as well as possible hypocentral migration along seismogenic zones and spatial extension of big-magnitude sources, these magnitudes are also assigned to the three nearest cells in each direction, as long as they fall within the seismogenic zone (always keeping the largest magnitude). This procedure defines locations and magnitudes of the seismic sources used to compute synthetic seismograms. The depth of the sources is set to 10 km. The receiver sites are situated over the area of interest on a $0.2° \times 0.2°$ grid displaced by $0.1° \times 0.1°$ from the grid of seismic sources. As a result, the shortest source to receiver distance is about 13.5 km. Synthetic seismograms are computed for all feasible source-site paths (taking the "site" structure along the whole propagation path). For thus defined source-site pairs, with a given FPS and magnitude, synthetic acceleration is computed for P-SV and SH components using the modal summation technique (PANZA, 1985; PANZA and SUHADOLC, 1987; FLORSCH et al., 1991). The summation is carried out up to the frequency of 1 Hz for which we can reasonably assume the point source approximation to be valid. Seismograms are computed for a unit seismic moment and then scaled to the magnitude of the earthquake using the magnitude-moment relation of KANAMORI (1977). The finiteness of the source is accounted for by scaling the spectrum using the spectral scaling law proposed by GUSEV (1983) as reported in AKI (1987). VACCARI (1995) has demonstrated that for periods between 1 and 2 s the Gusev spectral falloff produces higher spectral values than the omega squared spectral falloff, and thus generates more conservative results in our hazard computations.

Results

Maps of maximum ground velocity (Fig. 3) and maximum ground displacement (Fig. 4) for frequencies up to 1 Hz are compiled. A map of design ground acceleration (Fig. 5) is compiled by fitting the long period portion of the computed spectra to the response spectrum of Eurocode-8 (CEN, 1994) soil type A (rock and stiff deposits). The maximum values are to be found in the Friuli region in NE Italy and in western Slovenia. The sensitivity study has shown that the choice of the fault plane solution can influence the hazard by a factor of about 2 (e.g., strike slip vs. thrust event in the same seismogenic zone). For the city of Ljubljana DGA values vary by only 20% due to different FPS but change fourfold depending on the alternative selection of the location of 1511 event. This so called "Idrija" earthquake of March 26, 1511, almost completely controls the seismic hazard in this region. Its magnitude is estimated from macroseismic data to be 6.8 although its epicentre is still rather uncertain (RIBARIČ, 1979) since it may have occurred either in Friuli, in the Southern Alps or in the western part of the External Dinarides. The values of the hazard parameters gradually diminish eastwards, their lowest values being found in the northeastern part of Slovenia. This distribution of hazard values

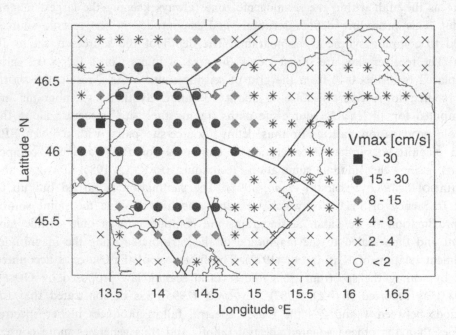

Figure 3
Map of maximum horizontal ground velocity (in cm/s) for frequencies below 1 Hz.

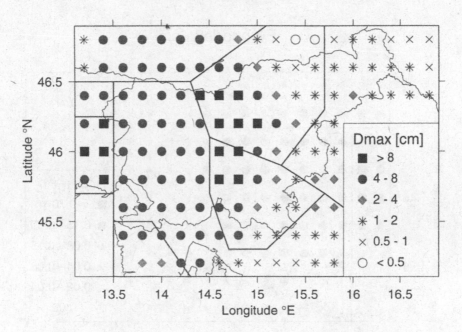

Figure 4
Map of maximum horizontal ground displacement (in cm) for frequencies below 1 Hz.

is rather similar to the one based on intensity data (RIBARIČ, 1986). However, the different levels of seismic hazard between western and eastern Slovenia are more pronounced in our study and compare well with the deterministic studies for the neighbouring regions.

To evaluate the reliability of the applied method, we have compared synthetics with the accelerograms of two earthquakes with epicentre in southern Slovenia recorded on free field strong motion instrument in Ljubljana, 50 km north of the epicentre. The synthetics were computed using published fault plane solutions (HERAK *et al.*, 1995) that are quite similar to the FPS proposed for this seismogenic zone. The values of the peak horizontal acceleration in the frequency band below 1 Hz, for which theoretical computations have been made and hazard parameters are mapped, are given in Table 1.

The peak value of acceleration is predicted within a factor of two. The acceleration response spectra for two events shown in Figure 6 match quite well the observed ones up to periods of about 4 s. For longer periods the noise dominates the records and no comparison is possible.

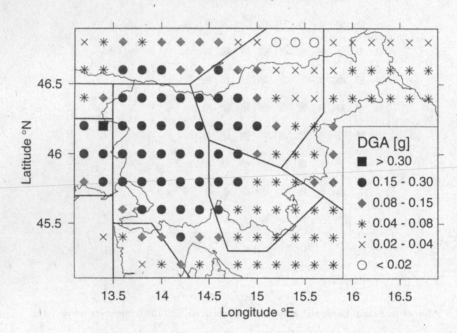

Figure 5
Map of design ground acceleration (in cm/s²) for Eurocode-8 soil type A.

Conclusions

The outlined deterministic method based on the computation of complete synthetic seismograms is able to delineate the main features of seismic hazard even for regions as small as Slovenia. Seismic hazard is controlled mainly by the

Table 1

Observed and synthetic peak ground accelerations for frequencies below 1 Hz for two earthquakes at 50 km distance

Event	ML	N–S observed PGA [g]	N–S synthetic PGA [g]	E–W observed PGA [g]	E–W synthetic PGA [g]
22.05.1995. 11:16	4.4	$1.2*10^{-5}$	$2.8*10^{-5}$	$0.9*10^{-5}$	$2.1*10^{-5}$
22.05.1995. 12:50	4.7	$1.5*10^{-5}$	$3.5*10^{-5}$	$1.2*10^{-5}$	$1.1*10^{-5}$

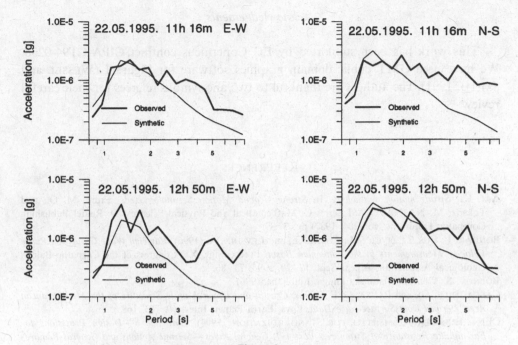

Figure 6

Comparison of observed and synthetic horizontal acceleration response spectra for frequencies below 1 Hz for two earthquakes at 50 km distance.

strongest event in the area and is found to be highest in western Slovenia, where the design ground acceleration reaches values exceeding 0.25 g.

Since its effects on seismic hazard determination are enormous, the main efforts in the future should be concentrated to precisely determine the size and location of the biggest event in the Slovenian catalogue, the Idrija event of March 26, 1511.

Peak values of acceleration in the case of Mt. Snežnik events are well modelled even with this rather rough approach, but the duration of the strong shaking is underestimated, partly because of the used point source approximation, partly because soft shallow sediments are not included in the modelling. However, if estimates of site effects for a given location are available, one can easily superimpose their effect to the "bedrock" motion estimated in this paper. Finally, we have shown that comparisons of the computed and recorded acceleration response spectra in the case of 1995 Mt. Snežnik exhibit considerable similarity, illustrating the appropriateness of the methodology applied in the paper.

Acknowledgements

This work has been supported by EC Copernicus contract CIPA-CT94-0238. We have used GMT public domain graphics software for Figure 1 (WESSEL and SMITH, 1991). The authors are thankful to two anonymous referees for their careful review.

REFERENCES

AKI, K., *Strong motion seismology*. In *Strong Ground Motion Seismology* (eds. Erdik, M. O., and Toksöz, M. N.) (NATO ASI Series C: Mathematical and Physical Sciences, D. Reidel Publishing Company, Dordrecht, vol. 204, 1987) pp. 3–39.

BONDÁR, I., BUS, Z., ŽIVČIĆ, M., COSTA, G., and LEVSHIN, A. (1996), *Rayleigh Wave Group and Phase Velocity Measurements in the Pannonian Basin*, Proceedings, XV Congress of the Carpatho-Balcan Geological Association, Athens Sept. 17–20, 1995, 73–86.

BUBNOV, S., *Potresi* (Mladinska knjiga, Ljubljana 1996).

CALCAGNILE, G., and PANZA, G. F. (1990), *Crustal and Upper Mantle Structure of the Mediterranean Area Derived from Surface-wave Data*, Phys. Earth Planet. Inter. *60*, 163–168.

CEN—EUROPEAN COMMITTEE FOR STANDARDIZATION (1994), *Eurocode 8—Design Provisions for Earthquake Resistance of Structures. Part 1-1: General Rules—Seismic Actions and General Requirements for Structures, European Prestandard, ENV 1988-1-1*.

CORNELL, C. A. (1968), *Engineering Seismic Risk Analysis*, Bull. Seismol. Soc. Am *58*, 1583–1606.

COSTA, G., PANZA, G. F., SUHADOLC, P., and VACCARI, F. (1993), *Zoning of the Italian Territory in Term of Expected Peak Ground Acceleration Derived from Complete Synthetic Seismograms*, J. Appl. Geophys. *30*, 149–160.

CVIJANOVIĆ, D. (ed.) (1974), *Map of distribution of maximum macroseismic intensities*. In *Proceedings of the Seminar of the Seismotectonic Map of the Balkan Region* (eds. Karnik, V., and Gorškov, G. P.) UNESCO-UNDP Project REM/70/172, Skopje. Appendix: Maps.

CVIJANOVIĆ, D., *Seizmičnost područja SR Hrvatske*, Disertacija (Sveučilište u Zagrebu, PMF, Zagreb 1981).

FIEGWEIL, E. (1981), *Austrian Earthquake Catalog* (Computer File), Central Institute of Meteorology and Geodynamics, Vienna.

FLORSCH, N., FÄH, D., SUHADOLC, P., and PANZA, G. F. (1991), *Complete Synthetic Seismograms for High-frequency Multimode Love Waves*, Pure appl. geophys. *136*, 529–560.

FRANKEL, A. (1995), *Mapping Seismic Hazard in the Central and Eastern United States*, Seismol. Res. Lett. *66* (4), 8–21.

GUSEV, A. A. (1983), *Descriptive Statistical Model of Earthquake Source Radiation and its Application to an Estimation of Short Period Strong Motion*, Geophys. J. R. Astron. Soc. *74*, 787–808.

HERAK, M., HERAK, D., and MARKUŠIĆ, S. (1995), *Fault Plane Solutions for Earthquakes (1956–1995) in Croatia and Neighbouring Regions*, Geofizika *12*, 43–56.

ISC (1976), *Bulletin of the International Seismological Centre*, May 1976.

JOKSOVIĆ, P., and ANDRIĆ, B., *Ispitivanje građe zemljine kore metodom dubokog seizmičkog sondiranja na profilu Pula—Maribor* (Geofizika, Zagreb 1983) 33 pp. (in Croatian, unpublished).

KANAMORI, H. (1977), *The Energy Release in Great Earthquakes*, J. Geophys. Res. *82*, 2981–2987.

KARNÍK, V. (1968), *Seismicity of the European Area, Part 1*, Academia, Czechoslovak Academy of Sciences, Praha.

KARNÍK, V. (1971), *Seismicity of the European Area, Part 2*, Academia, Czechoslovak Academy of Sciences, Praha.

KENNETT, B. L. N., and ENGDAHL, E. R. (1991), *Travel Times for Global Earthquake Location and Phase Identification*, Geophys. J. Int. *105*, 429–466.

KRINITZSKY, E. L. (1995), *Deterministic Versus Probabilistic Seismic Hazard Analysis for Critical Structures*, Eng. Geol. *40*, 1–7.

LAPAJNE, J. K., ŠKET MOTNIKAR, B., ZABUKOVEC, B., and ZUPANČIČ, P. (1997a), *Spatially Smoothed Seismicity Modeling of Seismic Hazard in Slovenia*, J. Seismol. *1*, 73–85.

LAPAJNE, J. K., ŠKET MOTNIKAR, B., and ZUPANČIČ, P. (1997b), *Preliminary Seismic Hazard Maps of Slovenia*, Natural Hazards *14*, 155–164.

LAPAJNE, J. K., ŠKET MOTNIKAR, B., and ZUPANČIČ, P. (1995), *Delineation of Seismic Hazard Areas in Slovenia*, Proceedings of the Fifth International Conference on Seismic Zonation, Nica *429*, 436.

MICHELINI, A., ŽIVČIČ, M., and SUHADOLC, P. (1998), *Simultaneous Inversion for Velocity Structure and Hypocenters in Slovenia*, J. Seismol. *2*, 257–265.

NESTEROV, A. N., and YANOVSKAYA, T. B. (1988), *Lateral Lithosphere Inhomogeneities in Southeastern Europe from Surface Wave Observations*, Izv. AN SSR, Fizika Zemli, *11*, 3–15 (in Russian).

NESTEROV, A. N. and YANOVSKAYA, T. B. (1991) *Inferences on Lithospheric Structure in Southeastern Europe from Surface Wave Observations*, XXII General Assembly ESC, Proceedings and Activity Report 1988–1990, *I*, 93–98.

OSSERVATORIO GEOFISICO SPERIMENTALE, OGS (1992), *ALPOR-Earthquake Catalogue for the Eastern Alps Region* (Computer File), Trieste.

PANZA, G. F., MUELLER, ST., and CALCAGNILE, G. (1980), *The Gross Features of the Lithosphere-asthenosphere System in Europe from Seismic Surface Waves and Body Waves*, Pure appl. geophys. *118*, 1209–1213.

PANZA, G. F. (1985), *Synthetic Seismograms: The Rayleigh Waves Modal Summation*, J. Geophysics *58*, 125–145.

PANZA, G. F., and SUHADOLC, P., *Complete strong motion synthetics*. In *Seismic Strong Motion Synthetics, Computational Techniques 4* (ed. Bolt, B. A.) (Academic Press, Orlando 1987) pp. 153–204.

PANZA, G. F., VACCARI, F., COSTA, G., SUHADOLC, P., and FAEH, D. (1996), *Seismic Input Modelling for Zoning and Microzoning*, Earthquake Spectra *12*, 529–566.

POLJAK, M., ŽIVČIČ, M., and ZUPANČIČ, P. (2000) *Seismotectonic Characteristics of Slovenia*, Pure appl. geophys. *157*, 37–55.

RAVNIK, D., RAJVER, D., POLJAK, M., and ŽIVČIČ, M. (1995), *Overview of the Geothermal Field of Slovenia in the Area between the Alps, the Dinarides and the Pannonian Basin*, Tectonophysics *250*, 135–149.

RIBARIČ, V. (1979), *The Idrija Earthquake of March 26, 1511*, Tectonophysics *53*, 315–324.

RIBARIČ, V. (1982), *Seismicity of Slovenia—Catalogue of Earthquakes (792 A.D.-1981)*, SZ SRS, Publication, Series A, *1-1*, Ljubljana, 650 pp.

RIBARIČ, V. (1986), *Prilozi proučavanju seizmičnosti i seizmičkog zoniranja Slovenije*, Disertacija, Prirodoslovno-matematički fakultet Sveučilišta u Zagrebu, 337 pp.

SHEBALIN, N. V., KARNÍK, V., and HADŽIEVSKI, D. (eds.) (1974), *Catalogue of Earthquakes. Part I 1901–1970; Part II prior to 1901*, UNDP/UNESCO Survey of the Seismicity of the Balkan Region. Skopje, 1–65 and Appendix.

SLEJKO, D., and KIJKO, A. (1991), *Seismic Hazard Assessment for the Main Seismogenic Zones in the Eastern Alps*, Tectonophysics *191*, 165–183.

SUHADOLC, P., and PANZA, G. F., *Physical properties of the lithosphere-astenosphere system in Europe from geophysical data*. In *The Lithosphere in Italy, Advances in Earth Science Research* (eds. Boriani, A., Bonafede, M., Piccardo, G. B., and Vai, G. B.) (Acad. Naz. Lincei 1989) pp. 15–44.

VACCARI, F. (1995), *LP-displacement hazard evaluation in Italy*. In *Proc. 24th General Assembly of the European Seismological Commission*, Athens, *3*, 1489–1498.

WESSEL, P., and SMITH, W. H. F. (1991), *Free Software Helps Map and Display Data*, EOS, Trans. Amer. Geophys. Un. *72* (441), 445–446.

WILLMORE, P. L. (ed.) (1979), *Manual of Seismological Observatory Practice*, World Data Center A for Solid Earth Geophysics, Report SE-20.

ZAJEDNICA ZA SEIZMOLOGIJU SFRJ (1987), *Tumač seizmološke karte SFR Jugoslavije*, Beograd, 1987, 6 pp.

ZSÍROS, T., MÓNUS, P., and TÓTH, L. (1988), *Hungarian Earthquake Catalog (456–1986)*, Seismological Observatory, Geodetic and Geophysical Research Institute, Hungarian Academy of Sciences, Budapest, 182 pp., Computer file (1992)—data from 456 to 1990.

ŽIVČIĆ, M., BONDÁR, I., and PANZA, G. F. (2000), *Upper Crustal Velocity Structure in Slovenia from Rayleigh Wave Dispersion*, Pure appl. geophys. *157*, 131–146.

(Received May 25, 1998, revised October 8, 1998, accepted October 8, 1998)

To access this journal online:
http://www.birkhauser.ch

Pure appl. geophys. 157 (2000) 185–204
0033–4553/00/020185–20 $ 1.50 + 0.20/0

© Birkhäuser Verlag, Basel, 2000

Pure and Applied Geophysics

A Contribution to Seismic Hazard Assessment in Croatia from Deterministic Modeling

S. MARKUŠIĆ,[1] P. SUHADOLC,[2] M. HERAK[1] and F. VACCARI[2]

Abstract—Some of the elements of regional seismic hazard in Croatia are assessed by computing synthetic accelerograms at a predetermined set of sites. The input dataset consists of structural models, parameters of seismic sources, and an updated earthquake catalog. Synthetic strong-motion time series for frequencies below 1 Hz are computed on a grid of sites using the modal summation technique. The long-period hazard is described by the distribution of estimated peak values of ground displacement, velocity and acceleration, while the short-period hazard is represented by the map of design ground acceleration values (DGA). The highest values of DGA exceeding 0.35 g on the base-rock level are found in the southeastern coastal part of the country, in the greater Dubrovnik area.

Key words: Seismic hazard, deterministic modeling, synthetic seismograms, Croatia.

1. Introduction

Traditionally seismic zoning has been one of the most important elements of post-earthquake surveying, proving essential in choosing which urban and non-urban areas are most safe or best suited to the development of new settlements. The recent guidelines of the United Nations sponsored International Decade for Natural Disaster Reduction (IDNDR) for the preparation of pre-catastrophe plans of action, enforce the idea that zoning and related seismic hazard assessment can and must be used as means of prevention in areas that have not yet been struck by disaster but are potentially prone to it.

The first scientific and technical methods developed for seismic zoning were deterministic and based on the observation that damage distribution is often correlated to the spatial distribution and physical properties of the underlying terrain and rocks (see REITER, 1990). The deterministic approach to earthquake hazard assessment as used here was developed by COSTA *et al.* (1993) for seismic zoning of the Italian territory, and was subsequently used for the zoning of Bulgaria (OROZOVA-STANISHKOVA *et al.*, 1996a) and Algeria (AOUDIA *et al.*,

[1] Department of Geophysics, Faculty of Sciences, University of Zagreb, Horvatovac bb, 10000 Zagreb, Croatia.
[2] Department of Earth Sciences, University of Trieste, Via E. Weiss 1, 34127 Trieste, Italy.

1998). The procedure is based on computation of complete synthetic seismograms using the modal summation technique (PANZA, 1985; FLORSCH *et al.*, 1991) on a predetermined grid of sites. The heterogeneity of the shallow earth structure in the region along with specific properties of individual seismic sources (or source zones) make reliable prediction of near-field waveforms using statistical methods an impossible task. The use of synthetic seismograms enables estimation of all engineering parameters needed to assess the seismic hazard, even in areas where scarce (or no) historical or instrumental information is available (e.g., AOUDIA *et al.*, 1998). Since the computation is transparent at each step, it is possible to evaluate the influence of each parameter on the final result. Moreover, the use of information concerning source mechanism and of available structural models of the region, by computing wave generation, propagation and attenuation, overcomes one of the major sources of uncertainty in probabilistic computations, i.e., the use of empirical attenuation laws, which are usually derived for quite different tectonic and structural settings and in most cases uncritically adopted. Deterministic methods also do not require catalog completeness, since only the strongest events (i.e., controlling earthquakes) of an area are considered. On the other hand no frequency of occurrence is taken into account. Deterministic hazard assessment results can be compared to probabilistic results only when these are computed for very large return periods.

Here we estimate the maximal displacements, velocities and accelerations in the Croatian and near-by territories that have been caused by previous earthquakes. Using the same procedure one could take into account the rate of earthquake occurrence within the specific seismogenic source zone (OROZOVA-STANISHKOVA *et al.*, 1996b) in order to obtain the ground-motion parameters corresponding to various risk levels and return periods.

2. *Input Data*

The computed synthetic seismograms are as good as is our knowledge of the source and propagation effects. Therefore, parameters describing seismic sources and structural models must be carefully defined and assigned to the studied area. The input data set consists of four main subsets. They are: i) structural models inside regional polygons, ii) seismogenic source zones, iii) characteristic fault plane solutions, and iv) the earthquake catalog (see the flow-chart scheme in Fig. 1). A brief description of each of them for the territory of Croatia is given below.

2.1 *Structural Models*

The structural models are defined within regional polygons separating areas characterized by different lithospheric properties. Structural models are represented

by a number of flat layers, each of which is described by its thickness, density, *P*-
and *S*-wave velocities and anelastic attenuation. Since the computation is aimed at
a first-order seismic zonation, the structural models do not explicitly account for
local site effects, and are representative of the average properties within each
polygon. In order to propose a suitable structural model, all available geophysical
and geological information for the investigated territory must be considered. Since
to date only the areas of the southern Adriatic and Dinarides have been investigated in some detail (HERAK, 1990; MARKUŠIĆ, 1991; HERAK and HERAK, 1995),
it was necessary to determine the crustal and uppermost mantle body-wave velocity
models for the entire region studied.

For this purpose, the Croatian territory is divided into eight regional polygons
(Fig. 2). Their borders are defined on the basis of specific geological properties and
the local topography of the Moho discontinuity. Here, the velocity models are

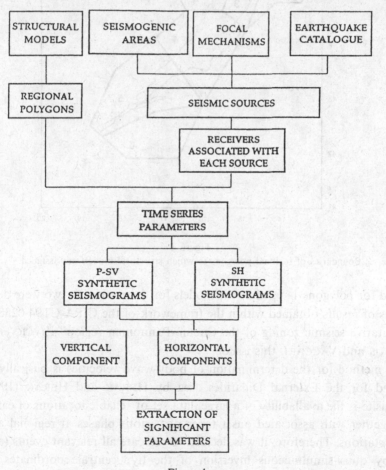

Figure 1
Flow-chart scheme of the seismic hazard assessment procedure used (after COSTA *et al.*, 1993).

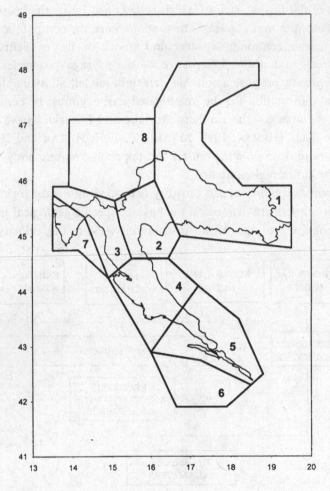

Figure 2
Boundaries of regional polygons to which structural models are assigned.

obtained for polygons 1–6, while the models for the remaining two were defined on the basis of results obtained within the framework of the CIPA-CT94-0238 project: "Quantitative seismic zoning of the Circum-Pannonian region" ŽIVČIĆ *et al.*, this issue; BUS and VACCARI, this issue).

The method for the determination of body-wave velocities is basically the one described for the External Dinarides area by HERAK and HERAK (1995). The prerequisite is the availability of a large data set of reliable locations of earthquake foci, together with associated onset times of various phases at regional and local seismic stations. Therefore, it was decided to relocate all relevant events (a total of 3434) by quasi-simultaneous inversion of the hypocentral coordinates and the so-called *inter-area* model parameters that control the shape of theoretical travel-time curves (see HERAK and HERAK, 1995, for details). The resulting inter-area models are rather simple (three layers with non-horizontal interfaces over a halfspace), each of them representing average properties of the crust and upper mantle between the chosen epicentral area and a particular group of seismic stations. Their purpose is to simulate lateral heterogeneity of the lithosphere in the area, enabling more accurate earthquake locations. The final locations were obtained after eight cycles involving earthquake location, model parameters, adjustments and station correction computations (for 598 phase-station pairs).

The final event locations, together with the corresponding phase onset times, constitute the input data set for *intra-area* velocity computations (the term *intra-area*—taken from HERAK and HERAK (1995)—denotes velocities *within* each of the regional polygons). The method employed considers all possible pairs of travel times observed at some stations for earthquakes within the selected polygon. If M stations reported arrival times for a total of K earthquakes, the velocity v in the epicentral region is estimated as

$$v = S_d/S_t,$$

where

$$S_d = \sum_{i=1}^{K} \sum_{j=1}^{K} \sum_{m=1}^{M} (d_{i,m} - d_{j,m}), \qquad S_t = \sum_{i=1}^{K} \sum_{j=1}^{K} \sum_{m=1}^{M} \cdot (t_{i,m} - t_{j,m})$$

are the sum of differences of raypaths $(d_{i,m})$ and travel times $(t_{i,m})$, respectively, for all possible pairs of earthquakes within each of the polygons. Indices i and j denote earthquakes, while m corresponds to stations. The method is fully described in HERAK and HERAK (1995). The original algorithm was slightly modified to allow the consideration of events within a given depth range, thus enabling estimation of the local velocity-depth profiles. When onset times of direct phases (such as P_g and S_g) are considered, only the velocities within the seismogenic layer (up to the depth of about 20 km in Croatia) may be estimated. The use of refracted waves (P_b, S_b, P_n and S_n) yields velocities representative of the uppermost parts of the lower crust and the upper mantle. For the upper crust, the velocities were estimated in a stack of overlapping layers whose thickness varied between 2 and 10 km, depending on the number of hypocenters located within each of them. Only the velocities with a standard error less than ± 0.03 km/s were accepted. In very few cases, when this threshold was exceeded, the velocity for this particular layer was obtained by linear interpolation. In the case of polygons 2–6, the values of computed body-wave velocities in the topmost layer were either too high (5.70–6.18 km s^{-1} for P-waves, probably due to dislocation of shallow foci), or there were not enough data to reliably estimate them. To obtain more realistic distribution of near-surface velocities, the uppermost layer in each of the mentioned polygons was subdivided into five thin layers. Velocities in these five layers were arbitrarily assumed as 60%, 75%, 90% and 100% of the values originally determined. Final velocity models for all regional polygons are presented in Table 1. It is assumed that mantle velocities below the last layer given in the table equal those in the IASP91 global model (KENNETT and ENGDAHL, 1991) while the density and quality factors are taken from the PREM model (DZIEWONSKI and ANDERSON, 1981). The densities in the crust are assumed to increase from 2500 kg/m^3 near the surface to 2800 kg/m^3 at the bottom of the crust. The models may be regarded as representative of the structure up to the base-rock level only. Modeling of the shallowest soil layers would require more detailed analyses on a considerably smaller scale than is the case here.

Table 1

Intra-area seismic velocity models within eight regional polygons in Croatia. α [km/s] and β [km/s] are P- and S-wave velocities in the layer of thickness h [km]. See text for details.

	Layers										
POLYGON 1											
α	3.10	5.20	5.90	6.80	7.84						
β	1.70	2.80	3.20	3.80	4.44						
h	3.0	2.0	9.0	12.8	∞						
POLYGON 2											
α	3.42	3.71	4.28	5.13	5.70	5.72	5.76	5.81	5.86	6.60	7.87
β	2.06	2.24	2.58	3.10	3.44	3.45	3.47	3.49	3.52	3.61	4.47
h	1.0	1.0	1.0	1.0	3.6	4.2	2.0	1.8	1.4	16.5	∞
POLYGON 3											
α	3.44	3.73	4.31	5.17	5.74	5.76	5.78	5.80	5.86	6.24	7.92
β	2.01	2.18	2.51	3.02	3.35	3.36	3.37	3.38	3.41	4.30	4.44
h	1.0	1.0	1.0	1.0	2.0	2.0	3.3	3.7	6.8	21.0	∞
POLYGON 4											
α	3.71	4.02	4.64	5.56	6.18	6.20	6.22	6.28	6.32	6.36	6.61	8.00
β	2.08	2.25	2.60	3.11	3.46	3.47	3.47	3.48	3.49	3.50	4.37	4.47
h	1.0	1.0	1.0	1.0	1.7	1.7	3.9	2.0	3.7	4.8	21.0	∞
POLYGON 5											
α	3.63	3.93	4.54	5.44	6.05	6.06	6.06	6.10	6.10	6.13	6.61	7.77
β	2.05	2.22	2.56	3.07	3.41	3.42	3.43	3.43	3.44	3.44	4.08	4.42
h	1.0	1.0	1.0	1.0	1.7	5.5	2.0	2.0	2.3	3.0	20.0	∞
POLYGON 6											
α	3.53	3.83	4.42	5.30	5.89	5.91	5.99	6.05	6.17	6.77	7.73
β	2.13	2.31	2.66	3.20	3.55	3.56	3.61	3.65	3.70	3.79	4.47
h	1.0	1.0	1.0	1.0	1.0	2.5	1.0	4.3	1.1	13.6	∞
POLYGON 7											
α	3.80	4.10	4.40	5.50	5.20	6.40	5.80	6.40	6.60	8.10	
β	2.10	2.35	2.50	2.90	3.00	3.60	3.40	3.60	3.75	4.60	
h	1.0	1.0	1.0	1.0	1.0	3.0	4.0	13.0	5.0	∞	
POLYGON 8											
α	3.50	4.50	5.00	5.20	5.50	5.80	6.00	6.40	7.20	8.00	
β	2.00	2.50	2.90	3.00	3.20	3.40	3.50	3.70	4.10	4.55	
h	0.1	0.3	1.6	4.0	4.0	4.0	5.0	10.0	10.0	∞	

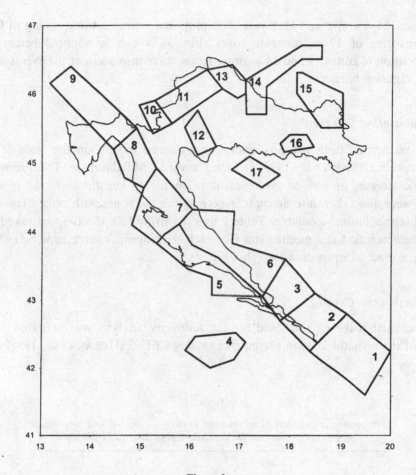

Figure 3
Seismogenic source zones in Croatia (after MARKUŠIĆ and HERAK, 1999).

Numerical experiments indicated that even small changes of assumed veloc-ities may have considerable influence on the final modeling. This is especially true for the topmost layers whose average properties are poorly known. The reliability of modeling will thus critically depend on the quality and representa-tiveness of structural models, which fact must be borne in mind when applying our results.

2.2 Seismogenic Source Zones

Delineation of seismogenic source zones is the final result of the seismotec-tonic zoning of a certain territory. Source zones are characterized by a specific tectonic, geodynamic and seismic behavior, which is assumed to be homogeneous within the zone. On the basis of local and regional seismicity, geology and tectonics, MARKUŠIĆ and HERAK (1999) proposed seismotectonic zoning of Croatia consisting of 17 seismogenic zones (Fig. 3), which is adopted herein with modification of zones 13 and 14 in order to also take into account the events across the Hungarian border.

2.3 Fault-plane Solutions

Characteristic fault plane solutions were defined mostly on the basis of data presented in HERAK *et al.* (1995) and the Harvard CMT database. The direction of tectonic stresses, as well as geological data on the strike, dip and rake of active faults were also taken into account, especially for the zones with only a few or no fault-plane solutions available. Table 2 lists the strike (φ), dip (δ) and rake (λ) of the characteristic focal mechanisms (for each seismogenic source zone SZ1–SZ17) that are used as input data (see also Fig. 4).

2.4 Earthquake Catalog

The earthquake catalog used for the seismicity analysis was prepared on the basis of the revised Croatian earthquake catalog CEC92 (HERAK *et al.*, 1996) which

Table 2

The parameters of fault plane solutions (strike—φ, dip—δ and rake—λ, in degrees) characteristic for each seismogenic zone (SZ1–SZ17).

	φ	δ	λ
SZ1	127	77	85
SZ2a	310	36	130
SZ2b	270	60	0
SZ3	310	58	114
SZ4	99	47	102
SZ5	286	61	22
SZ6	323	42	118
SZ7	310	46	120
SZ8	340	56	170
SZ9	336	84	180
SZ10	300	36	110
SZ11a	64	43	339
SZ11b	322	79	0
SZ12	120	56	100
SZ13	72	46	11
SZ14	64	43	339
SZ15	90	90	0
SZ16	267	55	22
SZ17	284	59	121

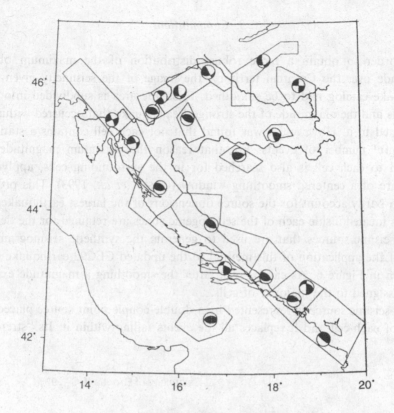

Figure 4
Characteristic fault-plane solutions within seismogenic zones.

has been supplemented with data for the years 1993–1996 (Fig. 5). It spans the period BC–1996, and may be considered complete for $M > 4$ for the entire 20th century. This threshold decreases in the last few decades of this century to about 3.6. The pre-instrumental part of the catalog has been compiled by many authors by searching and evaluating historical sources, and its completeness is difficult to assess. The rate of occurrence of large events ($M \geq 6.0$) estimated from the catalog is almost constant for the last 1000 years (1 event in approximately 30 years). We therefore believe that the catalog is reasonably complete at this level for the last millennium, which is adequate for deterministic hazard studies. The earthquakes have been assigned a local magnitude as listed in the catalog. In the case of events without magnitude, but with a reported intensity, the magnitude has been estimated using the magnitude-intensity-depth relation proposed by HERAK (1989). Furthermore, only main shocks shallower than 30 km were considered. The aftershocks were removed from the catalog by the algorithm suggested by KEILIS-BOROK et al. (1980).

3. Computations

In order to obtain a more robust distribution of the maximum observed magnitude over the Croatian territory, the image of the seismicity given by the earthquake catalog had to be smoothed. First, the area is subdivided into 0.2° × 0.2° cells and the magnitude of the strongest earthquake that occurred within a cell is assigned to it. However, it was found that not each cell contains a statistically meaningful number of events. For that reason the maximum magnitude to be assigned to each cell is also searched for in the surrounding cells, applying the procedure of a centered smoothing window (COSTA et al., 1993). This procedure can also partly account for the source dimensions of the largest earthquakes. Only the cells located inside each of the seismogenic zones are retained for the definition of the seismic sources that are used to generate the synthetic seismograms. The result of the application of this method to the updated CEC92 earthquake catalog is shown in Figure 6. As can be seen, after the smoothing a magnitude exceeding 5.5 is assigned to the majority of cells.

The seismic source is represented by a double-couple point source placed in the center of each cell and it replaces all the events falling within it. Its "strength" is

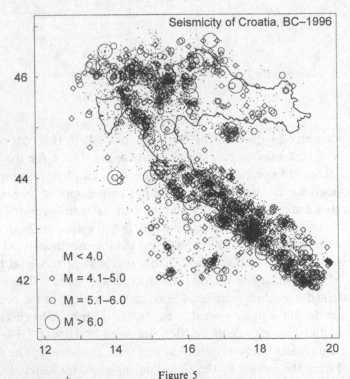

Figure 5
Epicenters of all earthquakes in Croatia and neighboring regions for the period BC–1996.

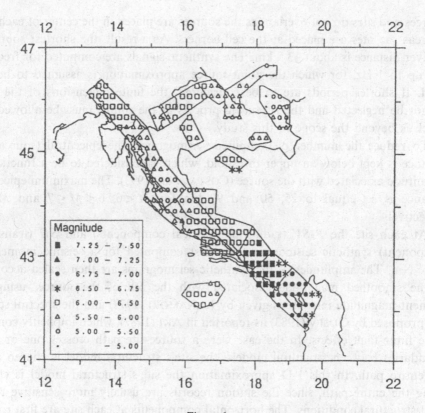

Figure 6
Representation of the "smoothed" seismicity within each seismogenic zone.

determined according to the maximum magnitude assigned to the cell. The orientation of the double-couple point source is made consistent with the predefined characteristic fault plane solution.

It is assumed that the hypocentral depth is variable, depending on the earthquake magnitude. For earthquakes with $M \leq 7.0$ the hypocentral depth is fixed at 10 km, while for those with $M \geq 7.0$ it is set equal to 15 km. This assumption is justified by the large errors affecting the hypocentral depth estimates and by the fact that strong ground motion is mainly controlled by shallow sources (e.g., VACCARI et al., 1990). The chosen depths of the sources are representative for the Croatian seismicity (HERAK and HERAK, 1990).

Once the structural models and the seismic sources are defined, sites are considered on a grid ($0.2° \times 0.2°$) covering the whole territory under investigation, and synthetic seismograms are computed at each of them by the modal summation technique (PANZA, 1985; FLORSCH et al., 1991). The modeling accounts not only for surface waves, but also for all those body waves which have phase velocities less than the S-wave velocity of the halfspace. For the velocity models used in our paper, this ensures the presence of the most energetic S-waves in the seismograms.

Sources and sites do not overlap, as the sources are placed in the center of each cell, whereas the sites are placed at the cell corners. As a result, the shortest source to receiver distance is about 13.5 km. The synthetic signals are computed for frequencies up to 1 Hz, for which the point-source approximation is assumed to be still valid. If shorter periods are to be considered, the finite dimensions of the fault cannot be neglected and the rupturing process at the source must be allowed for, which is beyond the scope of this study.

To reduce the number of computed seismograms, the epicentral (source-site) distance is kept below an upper threshold, which is considered to be a function of magnitude associated with the source (COSTA *et al.*, 1993). The maximum epicentral distance is set equal to 25, 50 and 90 km for $M < 6$, $6 \leq M \leq 7$ and $M > 7$, respectively.

At each site the *P-SV* (radial and vertical components) and *SH* (transverse component) synthetic seismograms are first computed for a seismic moment of 10^{-7} Nm. The amplitudes of the synthetic seismograms are then scaled according to the smoothed magnitude associated with the cell of the source, using the moment-magnitude relation as given by KANAMORI (1977) and the spectral scaling law proposed by GUSEV (1983) as reported in AKI (1987), which implicitly contains some finite-fault effects. In the case were a source-site path crosses one or more boundaries between structural models, the site's structural model is chosen along the entire path. In this 1-D approximation the site's structural model is chosen along the entire path, since the station records are usually more sensitive to the local structural conditions. The horizontal components at each site are first rotated into a reference system common to the whole territory (N–S and E–W directions) and then their vector sum is calculated. The resulting signal characterized by the largest amplitude due to any of the surrounding sources is selected and associated with that particular site.

The accelerograms obtained as outlined above are not suitable to be used for an overall earthquake risk estimation, as they only provide information on strong ground motion for periods above 1 *s*. They can, however, be used for estimation of the long-period risk, i.e., in those cases when one deals with objects whose natural periods lie above 1 *s*.

The complete set of computed accelerograms for one of the sites with the maximum peak value of ground acceleration (in the southeastern part of the investigated area, with the coordinates 43.0°N and 18.0°E) is shown as an example in Figure 7.

In order to extend the applicability of the computed set of synthetic accelerograms, the empirical knowledge on the spectral shapes of strong ground motion, observed globally or in the region under investigation, can be used to estimate the high-frequency components on the basis of long-period level and decay rate of the spectra. Here we have chosen to fit the response spectra computed for synthesized accelerograms to the long-period part of the theoretical spectra as given in

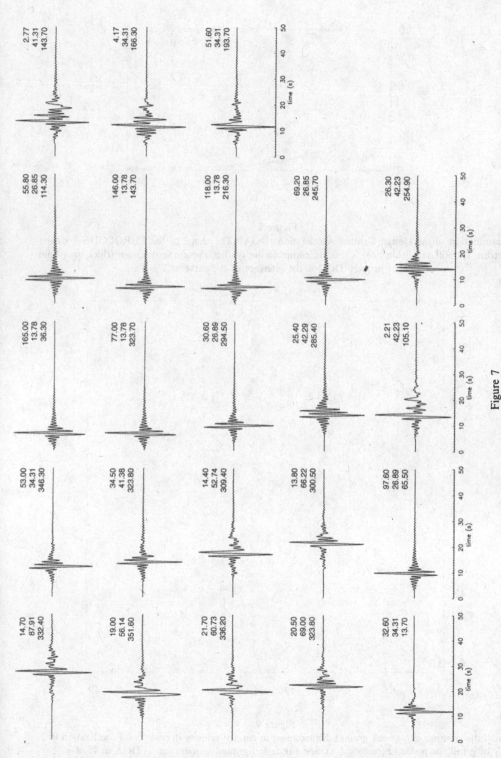

Figure 7

Complete set of accelerograms computed for the site (43.0°N, 18.0°E) with the highest peak ground acceleration within the studied region. The values reported in the right upper corner of each trace are (from top to bottom): peak acceleration (in cm/s²), epicentral distance (in km) and azimuth (in degrees) of the source-site line.

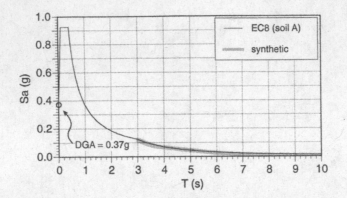

Figure 8
Determination of the Design Ground Acceleration (DGA). The shape of the EUROCODE 8 design spectrum for stiff soils (thin line) is used to complete the synthetic response spectrum (thick gray line) at short periods. DGA is the intercept of the curve at $T = 0$ s.

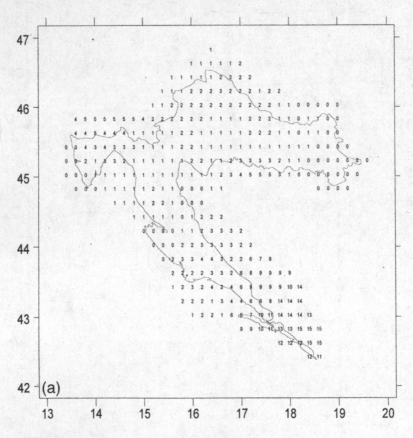

Figure 9
Spatial distribution of (a) peak ground displacement in cm, (b) velocity in cms^{-1}, (c) acceleration in % of g (all for periods exceeding 1 s), and (d) design ground acceleration — DGA in % of g.

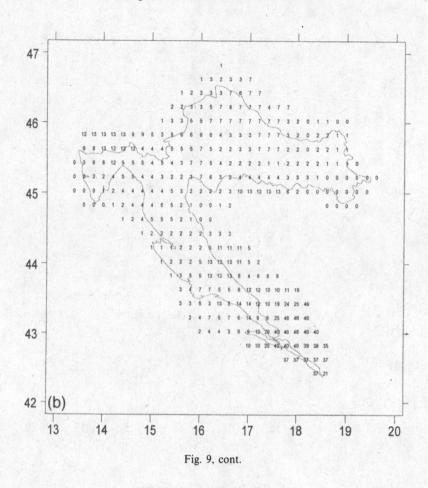

Fig. 9, cont.

EUROCODE 8 (1993). The period range in which the fit is searched is a function of the magnitude of the event (1–2 s for $M \leq 6$; 2–3 s for $6 < M \leq 7$; 3–5 s for $M > 7$). The example shown in Figure 8 refers to the strongest DGA obtained for Croatia, due to a $M = 7.4$ event, thus the fit is searched between $T = 3$ s and $T = 5$ s and the design spectrum is used to complete the synthetic one from $T = 3$ s to $T = 0$ s. In this way we define representative response spectra (RRS) for our set of site locations, and we may further define design ground acceleration (DGA) as the $T = 0$ s intercept of the corresponding RRS. The DGA is therefore the estimate of the peak acceleration (PGA) of the unfiltered accelerogram. The same procedure has been used by VACCARI et al. (1995) and PANZA et al. (1996), and has been validated empirically against recorded ground motions due to the Friuli 1976 and Irpinia 1980 earthquakes (PANZA et al., 1996).

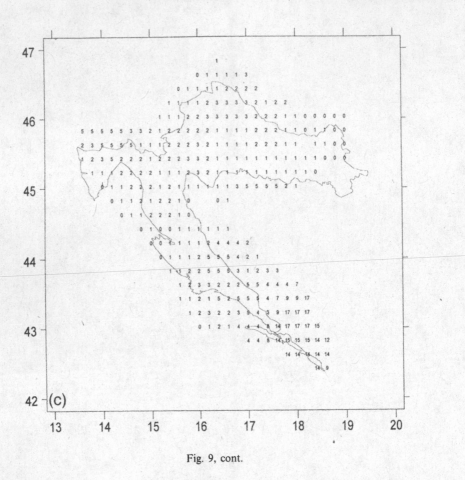

Fig. 9, cont.

4. Results and Conclusions

Spatial distribution of peak ground displacement, velocity and acceleration (for periods exceeding 1 *s*), and the design ground acceleration (DGA) are presented in Figures 9a–d. It is seen that all parameters are the largest in SE Dalmatia, with the highest values in the greater Dubrovnik area. The DGA there exceeds 0.35 g on the base-rock level. High DGA values are also estimated in northern Croatia (greater Zagreb region), in the southern part of the Pannonian Basin (due to seismicity of the Banja Luka area), and in the NW coastal parts of the country (Rijeka and northern Istria). The rest of the Croatian territory is characterized by DGA values less than 0.1 g on the base-rock level.

In order to compare the performance of the deterministic method as described here with the traditional technique, the peak horizontal ground acceleration is estimated by applying the empirical acceleration attenuation function GZ300 (PRELOGOVIĆ *et al.*, 1985) to the same set of sources and receivers as used above (Fig. 10). The GZ300 relation was derived for earthquakes in this region, and it

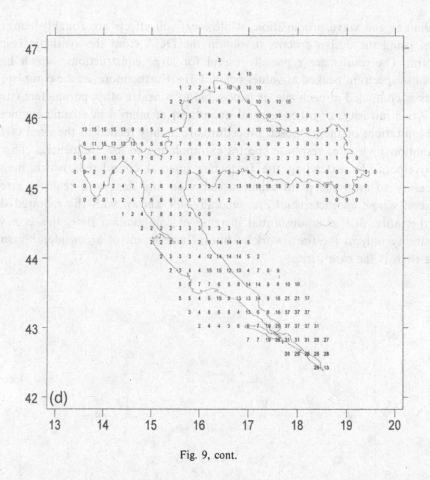

Fig. 9, cont.

yields expected maximal horizontal acceleration on average soil, given the magnitude and hypocentral distance. Comparison of Figure 9d and Figure 10 reveals that GZ300 yields higher values of peak acceleration than DGA by a factor of about 1.5–3 in most cases. This is due to local amplification effects, and is in agreement with the conclusion of MARKUŠIĆ (1997) that the GZ300 relation includes an average amplification factor of about 2 relative to the base-rock level. The discrepancies are mostly within the 1-sigma confidence limits of the GZ300 estimation (± 0.3 g for the logarithm of the estimated value). The major differences between GZ300 and DGA estimates are found for zones 5, 7 15 and 16 (Fig. 3). They are most probably caused either by possible errors in velocity models, or by the choice of the representative faulting mechanism (Fig. 4).

Computation of realistic synthetic seismograms for the Croatian territory yielded meaningful results, thus providing us with a powerful and economically valid scientific tool for seismic zonation and hazard assessment. The main advantage of the method lies in its ability to directly estimate the effects of source mechanics and wave propagation, while local soil effects are roughly considered when using the design spectra to obtain the DGA from the synthetic response spectra. The results are especially useful for large constructions, which have a response spectrum peaked at values below 1 Hz. Furthermore, as the complete time series is computed at each site, it is possible to consider other parameters (such as the Arias intensity, or other integral quantities) of interest in seismic engineering. The limitations of the proposed procedure are mainly related to the need of DGA estimation by way of response spectra extrapolation to low frequencies. This may be overcome by extending the frequency interval considered, which makes it necessary to drop the point-source assumption and deal with realistic extended sources, whose characteristics are generally not known with the required detail. Furthermore, due to a substantial increase of computation time, this is a viable possibility only in the framework of hazard assessment of a considerably smaller area than is the case here.

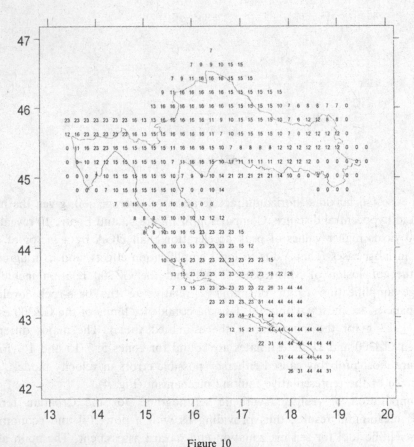

Figure 10
Peak horizontal accelerations in Croatia estimated using the empirical GZ300 relation (in % of g).

Acknowledgements

We wish to thank two anonymous reviewers for their constructive criticism of the manuscript. The study was supported by the Commission of European Communities Programme COPERNICUS "Quantitative Seismic Zoning of the Circum-Pannonian Region" (contract No. CIPA-CT94-0238), and by the Ministry of Science and Technology of the Republic of Croatia, grants No. 119298 and No. 119297.

REFERENCES

AKI, K., *Strong motion seismology.* In *Strong Ground Motion Seismology*, NATO ASI Series, Series C: Mathematical and Physical Sciences (eds. Erdik, M. Ö., and Toksöz, M. N.) (D. Reidel Pub. Co., Dordrecht 1987), *204*, 3–39.

AOUDIA, A., VACCARI, F., SUHADOLC, P., and MEGHRAOUI, M. (1998), *Seismogenic Potential and Earthquake Hazard Assessment in the Tell Atlas of Algeria*, J. Seismology, submitted.

BUS, Z., and VACCARI, F. (this issue), *Synthetic Seismogram Based on the Deterministic Estimation of Peak Ground Acceleration for the Hungarian Part of the Pannonian Basin*, this issue.

COSTA, G., PANZA, G. F., SUHADOLC, P., and VACCARI, F. (1993), *Zoning of the Italian Territory in Terms of Expected Peak Ground Acceleration Derived from Complete Synthetic Seismograms*, J. Appl. Geophys *30*, 149–160.

DZIEWONSKI, A. M., and ANDERSON, D. L. (1981), *Preliminary Reference Earth Model*, Phys. Earth Planet. Inter. *25*, 297–356.

EUROCODE 8 (1993), *Eurocode 8 Structures in Seismic Regions—Design—Part 1 General and Building*, Doc TC250/SC8/N57A.

FLORSCH, N., FÄH, D., SUHADOLC, P., and PANZA, G. F. (1991), *Complete Synthetic Seismograms for High-frequency Multimode Love Waves*, Pure appl. geophys. *136*, 529–560.

GUSEV, A. A. (1983), *Descriptive Statistical Model of Earthquake Source Radiation and its Application to an Estimation of Short-period Strong Motion*, Geophys. J. R. Astron. Soc. *74*, 787–800.

HERAK, D., and HERAK, M. (1990), *Focal Depth Distribution in the Dinara Mt. Region*, Gerlands Beiträge zur Geophysik *99*, 505–511.

HERAK, D., and HERAK, M. (1995), *Body-wave Velocities in the Circum-Adriatic Region*, Tectonophysics *241*, 121–141.

HERAK, M. (1989), *The Magnitude-intensity-focal Depth Relation for the Earthquakes in the Wider Dinara Region*, Geofizika *6*, 13–21.

HERAK, M. (1990), *Velocities of Body Waves in the Adriatic Region*, Boll. Geofis. Teor. Appl. XXXII *125*, 11–18.

HERAK, M., HERAK, D., and MARKUŠIĆ, S. (1995), *Fault-plane Solutions for Earthquakes (1956–1995) in Croatia and Neighbouring Regions*, Geofizika *12*, 43–56.

HERAK, M., HERAK, D., and MARKUŠIĆ, S. (1996), *Revision of the Earthquake Catalogue and Seismicity of Croatia, 1908–1992*, Terra Nova *8*, 86–94.

KANAMORI, H. (1977), *The Energy Release in Great Earthquakes*, J. Geophys. Res. *82*, 2981–2987.

KEILIS-BOROK, V. I., KNOPOFF, L., ROTWAIN, I. M., and SIDORENKO, T. M. (1980), *Bursts of Seismicity as Long-term Precursors of Strong Earthquakes*, J. Geophys. Res. *85*, 803–812.

KENNETT, B. L. N., and ENGDAHL, E. R. (1991), *Travel Times for Global Earthquake Location and Phase Identification*, Geophys. J. Int. *105*, 429–465.

MARKUŠIĆ, S. (1991), *Velocities of Refracted Longitudinal Pn Waves in the Dinarides Area*, Geofizika *8*, 101–113 (in Croatian with English abstract).

MARKUŠIĆ, S. (1997), *Deterministicko seizmicko zoniranje Hrvatske postupkom racunanja sintetickih seizmograma*, Ph.D. Thesis, University of Zagreb (in Croatian with English abstract), pp. 149.

MARKUŠIĆ, S., and HERAK, M. (1999), *Seismic Zonation of Croatia*, Natural Hazards *18*, 269–285.
OROZOVA-STANISHKOVA, I. M., COSTA, G., VACCARI, F., and SUHADOLC, P. (1996a), *Estimates of 1 Hz Maximum Acceleration in Bulgaria for Seismic Risk Reduction Purposes*, Tectonophysics *258*, 263–274.
OROZOVA-STANISHKOVA, I. M., VACCARI, F., and SUHADOLC, P. (1996b), *A new methodology for seismic hazard estimation*. In ESC Abstracts, XXV General Assembly, September 9–14, 1996, Reykjavik, Iceland.
PANZA, G. F. (1985), *Synthetic Seismograms: The Rayleigh Waves Modal Summation*, J. Geophys. *58*, 125–145.
PANZA, G. F., VACCARI, F., COSTA, G., SUHADOLC, P., and FÄH, D. (1996), *Seismic Input Modeling for Zoning and Microzoning*, Earthquake Spectra *12*, 529–566.
PRELOGOVIĆ, E., SKOKO, D., KUK, V., MARIĆ, K., MILOŠEVIĆ, A., ŽIVČIĆ, M., HERAK, M., ALLEGRETTI, I., and SOVIĆ, I. (1985), *Determination of the Earthquake Characteristics S_1 and S_2 on the Location of the Nuclear Plant Slavonija*, Faculty of Mining and Faculty of Science, Geophysical Institute, University of Zagreb, 130 pp. (in Croatian).
REITER, L., *Earthquake Hazard Analysis* (Columbia University Press, New York, 1990), 254 pp.
VACCARI, F., SUHADOLC, P., and PANZA, G. F. (1990), *Irpinia, Italy, 1980 Earthquake: Waveform Modelling of Strong Motion Data*, Geophys. J. Int. *101*, 631–647.
VACCARI, F., COSTA, G., SUHADOLC, P., and PANZA, G. F. (1995), *Zonazione sismica deterministica al prim'ordine per l'area italiana*. In *Terremoti in Italia: previsione e prevenzione dei danni*, Atti dei Convegni Lincei *122*, 117–126.
ŽIVČIĆ, M., BONDÁR, I., and PANZA, G. F. (2000), *Velocity Models of Slovenian Territory from Surface Wave Dispersion*, Pure appl. geophys. *157*, 131–146.

(Received April 25, 1998, revised November 2, 1998, accepted November 2, 1998)

To access this journal online:
http://www.birkhauser.ch

Pure appl. geophys. 157 (2000) 205–220
0033–4553/00/020205–16 $ 1.50 + 0.20/0

Pure and Applied Geophysics

Synthetic Seismogram Based Deterministic Seismic Zoning for the Hungarian Part of the Pannonian Basin

ZOLTÁN BUS,[1] GYÖZÖ SZEIDOVITZ[1] and FRANCO VACCARI[2]

Abstract—Deterministic seismic hazard computations have been done for the Hungarian part of the Pannonian basin within the framework of a cooperation of five countries. Synthetic seismograms have been computed by the modal summation method up to 1 Hz in order to determine the expected maximum displacement (D_{MAX}), velocity (V_{MAX}) and the design ground acceleration (DGA) on a $0.2° \times 0.2°$ grid. DGA values have been estimated from the seismograms by using the EUROCODE 8 (1993) standard.

This investigation justified the suspicion that a considerable part of seismic hazard of Hungary comes from the seismogenic zones of the neighbouring countries. The highest DGA reaches a value as high as 0.14 g (which corresponds approximately to the VIII intensity degree in the MSK-64 scale). Among the six largest cities of Hungary, three art particularly subject to a high seismic risk. Greater acceleration values have been found for the cities of Szeged and Debrecen than was expected before this study.

Key words: Synthetic seismograms, deterministic modelling, seismic hazard, design ground acceleration.

Introduction

In recent years it became obvious that seismic zoning can be a useful tool not only for the surveying of the areas struck by disaster, but for territories which were not affected by earthquakes in the written history.

In this study a new deterministic seismic zoning technique developed by COSTA et al. (1993) has been applied. This procedure has been used with success in regions with diverse geological endowments, e.g. Italy (COSTA et al., 1993), Algeria (AOUDIA et al., 1996) and Bulgaria (OROZOVA-STANISHKOVA et al., 1996).

The seismic hazard computations have been done for the Hungarian part of the Pannonian basin within the framework of a cooperation of five neighbouring countries (Croatia, Hungary, Italy, Romania and Slovenia).

[1] Seismological Observatory of GGRI of Hungarian Academy of Sciences, Meredek u. 18, H-1112 Budapest, Hungary, e-mail: bus@seismology.hu and szeido@seismology.hu
[2] CNR-Gruppo Nazionale per la Difesa dai Terremoti, Università di Trieste, Dipartimento di Scienze della Terra, via Weiss 4, I-34127, Trieste, Italy, e-mail: vaccari@geosun0.univ.trieste.it

The territory of Hungary is seldom impacted by severe, destructive earthquakes. Nevertheless from time to time (approx. every 10–20 years) damaging earthquakes occur. One of the most active seismic zones lies in the vicinity of Budapest, the capital of Hungary. The most significant Hungarian earthquake of the XX century (Dunaharaszti, 12 January 1956, see e.g. in SZEIDOVITZ, 1986b) arose only 20 km from the centre of the city. Although in the last four centuries no earthquake with magnitude $M > 5.5$ occurred in the very area of Budapest, with this new method it is possible to see the effects of such a hypothetical event.

Many of the large cities of the country with great economical and cultural significance lie in the neighbourhood of active seismogenic zones. It is important to know the level of seismic hazard in their area.

Figure 1 shows the geographical units of Hungary and the location of its largest cities together with the epicentres of the most damaging earthquakes (after ZSIROS *et al.*, 1988).

A considerable part of the seismic risk to Hungary originates in surrounding countries. International cooperation was important during this work, since specialists of the other countries have a detailed knowledge of the properties of source zones affecting the Hungarian settlements.

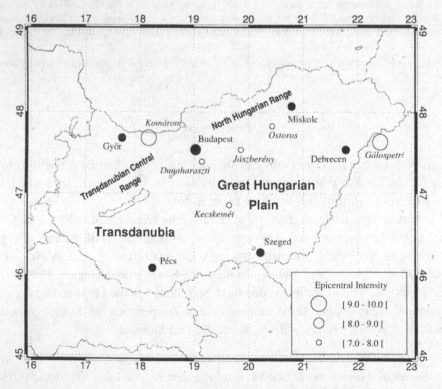

Figure 1
The geographical units of Hungary, its largest cities (filled circles) and the epicentres of the most damaging earthquakes (empty circles).

To date in Hungary mainly probabilistic seismic hazard estimations have been made. For example ZSIROS (1985) determined the maximum intensity values which will not be exceeded in 200 years with a 70% probability. Using a different approach, ZSIROS and MÓNUS (1984) computed a theoretical maximum intensity map together with acceleration, velocity and displacement data, using the intensity values of Hungarian earthquakes which occurred between 1859 and 1982. Additionally, detailed studies have been made to determine the properties of the most active seismogenic zones (compiling isoseist maps, recurrence-time analysis, seismic hazard estimation), see e.g., SZEIDOVITZ (1986a).

In the Soviet era, experts from the Soviet Union assessed qualitatively and quantitatively the level of seismic risk of the Eastern European states, see e.g., REISNER and SHOLPO (1975), although usually their estimates vary significantly from those of the given countries' scientists.

Tectonic and Seismicity of the Pannonian Basin

The Pannonian basin is a tectonically complex area which is encompassed by the Carpathian Mountains, the Dinarides and the Eastern and Southern Alps. In essence, the Pannonian basin is a set of small and rather deep subbasins, whose depth often exceeds 5 km, filled with Neogen-Quaternary sediments. Among the subbasins the ridges of the pretertiary basement can be found. Practically the whole territory of Hungary is underlain by the Pannonian basin.

The Pannonian basin and the surrounding mountains are the outcome of the collision of the European plate and the small plate fragments originating from the south which started in the Cretaceous. In the lower and middle Miocene the area within the Carpathian loop underwent an extensional period due to the subduction of the external Carpathian crust and the marginal part of the attenuated European continental crust (CSONTOS et al., 1992 and HORVÁTH, 1993). After the syn-rift period two compressional events took place (the latter is ongoing) and the general thermal subsidence of the entire basins system occurred between the (HORVÁTH, 1995).

The Pannonian basin is filled with thick Neogene and rather thin Quaternary sediment layers. The total thickness of the Neogene-Quaternary sediments in the subbasins is approximately 3 km, sometimes reaching a value of 8 km (STEGENA et al., 1975).

The Hungarian part of the Pannonian basin does not produce intense seismic activity. Because of the few large magnitude events and due to the lack of sufficient reliable focal solutions, it is not an easy task to make a connection between the faults determined by geologists and the observed earthquake epicentre distribution (GUTDEUTSCH and ARIC, 1988). In light of recent investigations by one of the authors (Gy. Szeidovitz), it is probable that a considerable part of the Hungarian earthquakes cannot be correlated with known faults, rather they are brought about by the effects of sedimentation in the basins.

The Most Severe Earthquakes of Hungary

The most destructive historical earthquakes of Hungary are listed in Table 1, where M is the local magnitude of the event, I_0 is the epicentral intensity and A_0 is the estimated acceleration in the epicentre. The relationship between intensity and acceleration described in BISZTRICSÁNY (1974) has been used throughout this paper. The events of unknown origin in the distant past have been omitted from the list, although they were quite strong in some cases. The epicentres of the listed events are shown in Figure 2 by empty circles.

Method

The method used for the deterministic seismic zoning of Hungary was developed at the Istituto di Geodesia e Geofisica (now Dipartimento di Scienze della Terra) of the University of Trieste, see COSTA *et al.* (1993) and PANZA *et al.* (1996).

The assessment of maximum ground acceleration, velocity and displacement is based on synthetic seismogram computation by the modal summation method (a detailed description can be found e.g., in PANZA, 1985 and FLORSCH *et al.*, 1991), which accounts for surface waves and for all those body waves which have phase velocities less than the S-wave velocity of the halfspace. In the case of our models, the most energetic S-waves are present in the seismograms. This technique makes it possible to efficiently model the wave propagation in a one-dimensional anelastic layered structure for arbitrary types of sources (e.g., explosion, double-couple, finite-length fault).

The parameters of the seismic sources and the structure in which the waves propagate are needed. The sources are confined to seismogenic areas, and the structure is simplified to a set of one-dimensional models.

Table 1

The most destructive earthquakes in Hungary (after SZEIDOVITZ, 1986b; SZEIDOVITZ and BUS, 1995 and ZSIROS et al., 1988)

Epicentre	Date	M	I_0	A_0 (g)
Érmellék (Gálospetri)	15 October 1834	6.5	9.0	0.2–0.4
Komárom	28 June 1763	6.2	9.0	0.2–0.4
Dunaharaszti	12 January 1956	5.6	7.5	0.075–0.1
Jászberény	21 June 1868	5.3	7.5	0.075–0.1
Eger (Ostoros)	31 January 1925	5.0	7.5	0.075–0.1
Kecskemét	8 July 1911	5.6	7.0	0.05–0.1

The seismogenic areas of the territory are defined on the basis of seismological and seismotectonic data. It is assumed during the computation that earthquakes occur only in these areas. For each seismogenic zone one or more representative focal plane solution is used.

The investigated area is divided into a system of flat layered structures. This subdivision is made on the grounds of the geological and geophysical attributes of the lithosphere under the investigated area.

The investigated territory is subdivided by a uniform grid (in our case the grid is 0.2° × 0.2°). The simulated sources and the focal mechanisms are assigned to the middle of the grid cells, while the receivers are located in the corners of the cells. For every cell a magnitude value is obtained by the use of a centered smoothing window. In this way the surrounding cells can affect the center one. The seismogenic zone's representative focal mechanism is assigned to each cell to which an event is assigned.

For the purpose of diminishing the number of synthetic seismogram computations, distance limits have been defined for specific magnitude ranges. As the hypocentre's depth is the most uncertain data among the earthquake location parameters, and the events with shallow sources are the most damaging, it is an

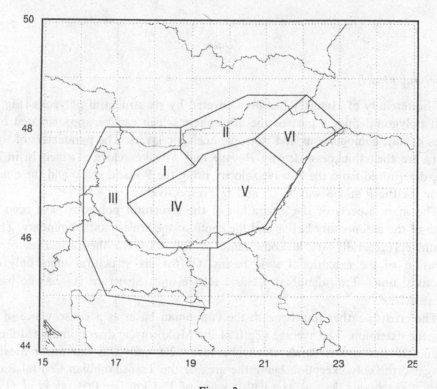

Figure 2
The boundaries of the structural units.

acceptable approximation to set all the hypocentres' depth to a fixed (in our case 10 km) value.

P–SV and *SH* seismograms are computed at each receiver for a seismic moment of 10^{-7} Nm, then the synthetic seismograms are scaled using the spectral scaling law proposed by GUSEV (1983) as reported in AKI (1987). If the propagation path crosses the structural units' boundaries, the structure at the receiver is used in the computation as the representative of the whole path. Having determined the *P–SV* and *SH* seismograms for all possible sources associated with a receiver, the peak of the signal with the largest horizontal amplitude is chosen for the representation of ground-shaking on the map.

The synthetic seismograms are computed to an upper frequency limit of 1 Hz, so the point-source approximation can be considered valid. In the case of higher frequencies, the dimension of the fault and the rupturing process should be taken into account.

The frequencies lower than 1 Hz are important from a practical point of view as the multi-story buildings, bridges, etc. which are common in large cities have the peak response frequency in this range (CSÁK *et al.*, 1981).

Data

Structural Models

The territory of Hungary has been covered by six structural polygons (Fig. 2). Each polygon defines a part of the lithosphere which can be approximated by a series of flat, homogeneous and isotropic anelastic layers. The parameters of these layers are their thickness, density *P*-wave and *S*-wave velocity (which is, in our case, determined from the *P*-wave velocity using the Poisson ratio) and the quality factor for the *P* and *S* waves, Q_α and Q_β, respectively.

The main aspects of the separation of the structural polygons have been the depth of the sedimentary basins and the depth of the crust-mantle boundary. These parameters generally are in coincidence as follows from the properties of the evolution of the extensional style basins. Of the six structural units only one structural unit—Transdanubian Central Range unit (Structure I)—has no basin-like structure.

The crust is rather thin beneath the Pannonian basin as a consequence of the Miocene extension. The average depth of the Mohorovičić discontinuity (Moho) is around 26 km and it varies slowly between 22.5 and 27.5 km under most of Hungary, the only exception being the area of the Transdanubian Central Range where it reaches its deepest part with a value of 32.5 km, see POSGAY *et al.* (1986) and HORVÁTH (1993).

The depth of the pretertiary basement under Hungary has been taken from the map of KILÉNYI and SEFARA (1989), the thickness of the Neogene-Quaternary sediments from the paper of STEGENA *et al.* (1975).

The velocity data stems from three sources. For the velocity and density values inside the basins we have used the data of SZABÓ and PÁNCSICS (1994). The primary source for the crustal velocities has been the paper of BONDÁR *et al.* (1996), who has made single-station group velocity measurements and a genetic algorithm based inversion for the area of the Pannonian basin. His results have been checked and amended by the data of MÓNUS (personal communication, 1998), who has determined a one-dimensional, two-layered crustal model by evaluating several hundred seismograms recorded in the region. Below the crust, the velocities of the IASPEI91 model (KENNETT and ENGDAHL, 1991) have been used.

Seismogenic Zones

We have defined fourteen seismogenic zones for Hungary and for those parts of the neighbouring countries close to the Hungarian boundary (Fig. 3). As can be seen in the figure, there are gaps between the zones, in other words there is no

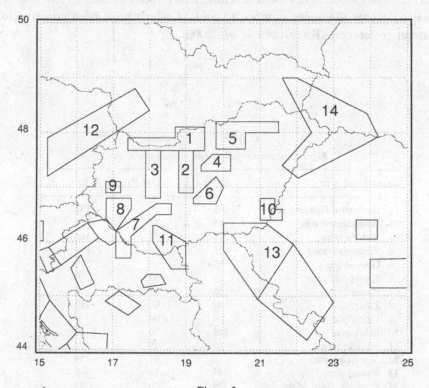

Figure 3
The boundaries of the seismogenic zones (their names are declared in Table 2).

"background zone" provided. This is because we are only considering earthquakes with magnitude 5 or above and there are no such events occurring outside the seismogenic zones we defined. There are also no active faults that could generate $M > 5$ events known. Therefore we apply the procedure strictly as it has been applied in the original paper by COSTA *et al.* (1993), to produce results consistent with what has already been computed in Italy and other countries. It is in any case possible to make different hypotheses about seismicity, to perform the computations again and to compare the results. Parametric analyses are easy to perform in this deterministic approach, nonetheless they are beyond the purpose of this paper.

The definition of zones is based on the seismic lineaments determined from the distribution of the earthquake epicenters (ZSIROS *et al.*, 1988) and the properties (amount, strike, type extent) of the known faults in the area (BARVITZ *et al.*, 1990).

In some cases the delineation of seismogenic zones boundaries was quite straightforward, based on earthquake data and the knowledge of the areas' tectonical settings. In other cases it was somewhat subjective due to the sparse seismicity or complicated tectonics of the given area.

The relationship between the numbering of the seismogenic zones and their names used in the text can be found in Table 2.

All the seismogenic zones listed in Table 2 have been defined by the authors of this paper, except the zones number 13 and 14 which have been constructed by Romanian researchers (RADULIAN *et al.*, 2000).

Table 2

Focal mechanisms assigned to seismogenic zones

Seismogenic zone	Strike (°)	Dip (°)	Rake (°)
1. **Hurbanovo-Diósjenö zone**	**261**	**29**	**−132**
2. **Dunaharaszti zone**	**85**	**73**	**159**
3. **Berhida zone**	**227**	**77**	**−166**
4. **Jászberény zone**	**125**	**66**	**137**
5. Ostoros zone	60	45	270
6. Kecskemét zone	65	90	0
7. Kapos zone	65	90	0
8. Zala zone	90	90	180
9. Ukk-Türje zone	70	90	0
10. **Békés zone**	**266**	**70**	**42**
11. Mecsek zone	80	45	90
12. **Mur-Mürz zone**	**0**	**61**	**158**
13. **Bánság zone**	**103**	**72**	**−5**
14. Érmellék (Satu Mare) zone	329	86	−42

Fault Plane Solutions

The fault plane solutions (FPS) emanate from the paper of GERNER (1995) who collected and re-evaluated the FPSs of the Pannonian basin published in the last few decades.

However there have been seven seismogenic zones for which it was impossible to find any FPS computed from earthquake data. In these cases the strike and type of the faults have been determined with the aid of the tectonic map of Hungary by BARVITZ *et al.* (1990). The small magnitude seismicity of the country makes it difficult to get FPSs, as there are only a few events in the Pannonian basin which are large enough to ensure a sufficient number of observations for accurate FPS determination.

The strike, dip and rake angles are shown in Table 2. When more than one FPS belonged to a seismogenic zone, the one with greater magnitude was chosen, as it is considered more reliable. The rows of Table 2 are typed with bold letters if they are the results of FPS computation from earthquake data.

Earthquake Catalogue

The earthquake catalogue used for the computation (ZSIROS *et al.*, 1988) covers the years between 456 and 1986. This database was completed up to the year 1989. Figure 4 shows the magnitude distribution in the seismogenic zones after the smearing. The macroseismic magnitude has been estimated from the maximum intensity (I_0) and the focal depth (h), using the Gutenberg-Richter formula:

$$M = 0.6\, I_0 + 1.8 \log h - 1.0. \tag{1}$$

In the case of unknown focal depth a value of 10 km was assumed. The only considered events in the computations were those whose errors in the location of epicentres are smaller than 20 km (with this restriction our data come from the years ranging from 1443–1989).

Results

Figures 5–7 show the output of the computations for the design ground acceleration (DGA), the maximum values of displacement (D_{MAX}) and velocity (V_{MAX}), respectively.

Design Ground Acceleration

As the frequency content of the synthetic seismograms limited to 1 Hz, the maximum values of acceleration (A_{MAX}) extracted from the synthetic seismograms

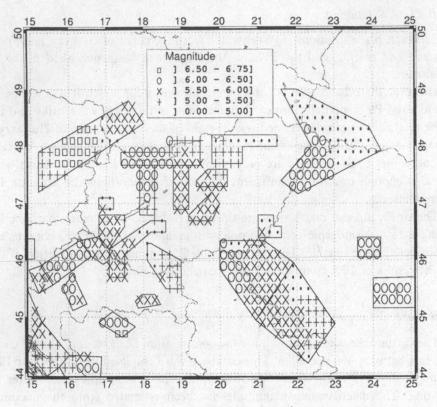

Figure 4
The seismogenic zones with the smeared magnitude distribution.

underestimate the true peak of acceleration. To overcome this limitation we may extend the deterministic results to higher frequencies by using the design response spectra (PANZA *et al.*, 1996), for instance EUROCODE 8 (1993), which define the normalized elastic acceleration response spectrum of the ground motion.

In general, this operation should be made taking into account the soil type. The structural models used in our computations are all of type A, as defined in Eurocode 8: "*stiff deposits of sand, gravel or overconsolidated clay, up to several tens of m thick, characterized by a gradual increase of the mechanical properties with depth (and by v_s values of at least 400 m/s at a depth of 10 m)*". The well data of SZABÓ and PÁNCSICS (1994) indicate that this assumption is reasonable for the territory of Hungary.

Therefore we can determine the Design Ground Acceleration (DGA) by fitting the response spectra computed from the synthetic seismograms (in the period range between 1 s and 5 s) with the one given by Eurocode 8.

Validation of the use of DGA instead of PGA (peak ground acceleration) can be found for accelerograms of the Italian Irpinia (1980) and Friuli (1976) earthquakes in VACCARI *et al.* (1995) and PANZA *et al.* (1996).

The distribution of DGA (Fig. 5) clearly shows that the regions with the highest seismic risk can be found near the borders of the country. The maximum values of DGA in these areas are between 0.08 and 0.15 g. In the MSK-64 scale, intensity degrees VII and VIII correspond to these acceleration values.

A significant region of the Great Hungarian Plain has no remarkable seismic risk, and this is the same for a region at the southern Hungarian part of the Danube. Due to this computation most of the sources inside Hungary do not cause accelerations higher than 0.08 g (approximately intensity VII).

The results show that we can expect epicentral intensity VI for the Kecskemét and Dunaharaszti, and an intensity value of VII for the Berhida source zone. The epicentral intensity experienced for the Kecskemét earthquake was VII (SZEIDOVITZ and BUS, 1995), for the Dunaharaszti earthquake it was between VII and VIII (SZEIDOVITZ, 1986b) and for the Berhida event it was VI (TÓTH et al., 1989). We can see there is approximately a 1.0–1.5 degree difference between the experienced and theoretical epicentral intensity values. We will discuss the causes of this phenomenon later.

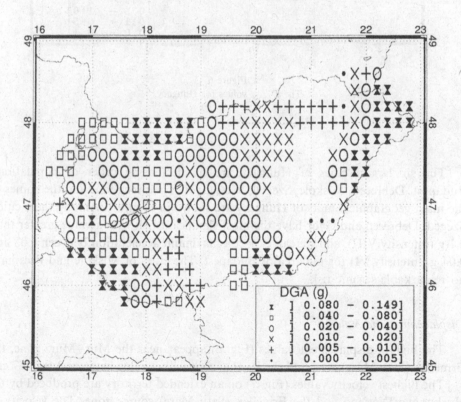

Figure 5
The DGA values for Hungary.

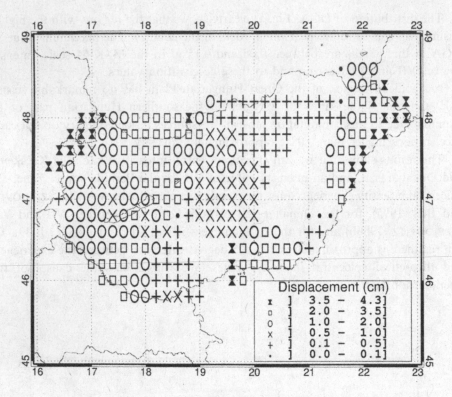

Figure 6
The D_{MAX} values for Hungary.

The six largest cities of Hungary are (in decreasing order of population): Budapest, Debrecen, Miskolc, Szeged, Pécs, Györ. Their locations and the names of the main geographical units of Hungary are shown in Figure 1. Among these cities, Szeged, Debrecen and Györ have the most significant risk with a value greater than 0.1 g (intensity VIII), and we can expect a maximum acceleration between 0.02 and 0.04 g (intensity VI) for the area of Budapest. The cities of Miskolc and Pécs have no remarkable seismic risk.

Displacement and Velocity

The highest displacement values (Fig. 6) appear near the Mur–Mürz zone, the Érmellék (Satu Mare) zone and the Bánság zone with a value between 3.5 and 7 cm.

The highest velocity values (Fig. 7) on an extended territory are produced by the Hurbanovo-Diósjenö and the Érmellék (Satu Mare) source zone. The velocity in these regions is between 8 and 15 cm/s.

Conclusions

Synthetic seismogram based deterministic seismic hazard computations have been performed for the territory of Hungary within the framework of international cooperation. The results show that we can expect considerable seismic hazard in three of the six largest cities of Hungary, namely the design ground acceleration reaches and passes the value of 0.1 g which corresponds roughly to VIII degree intensity on the MSK-64 scale. For a significant part of the Great Hungarian Plain the seismic hazard can practically be neglected.

As a matter of fact, the seismic hazard of Hungary originates mainly from abroad. Its main sources are the Mur–Mürz zone (Austria), Medvednica–Kalnik zone (Croatia), Bánság zone (Romania–Yugoslavia) and the Érmellék (Satu Mare) zone (Romania).

As we have seen, some of the results obtained in this paper—chiefly in the case of intra-Hungarian earthquakes—contradict the macroseismic observations. A possible source of the discrepancy is that the investigated territory can be characterized with laterally highly heterogeneous thin sediments with very low wave veloc-

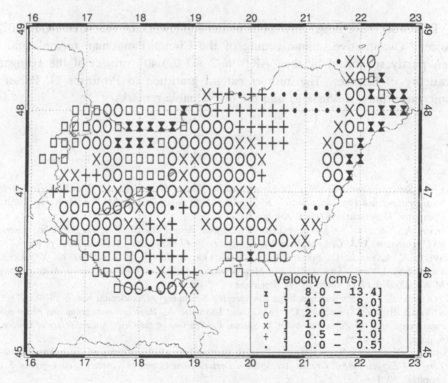

Figure 7
The V_{MAX} values for Hungary.

ities due to the before-mentioned flavours of the Pannonian basin. These hetero-geneities can be handled efficiently only by two-dimensional methods. Experience demonstrates that these site effects can change the intensity values in some cases by two degrees on the MSK-64 scale. As the aim of this paper was the first-order seismic zoning of Hungary, the investigation of local site effects is beyond the scope of this work.

We have compared our results for the largest cities with those of ZSIROS (1985), who determined the intensities which will not be exceeded in 200 years with a 70% probability. We have found that our estimates are the same for Budapest and Miskolc, Zsiros gained higher intensity for Pécs (VI vs. V), we have higher intensity for Győr (VIII vs. VII), Debrecen (VIII vs. V) and Szeged (VIII vs. V).

Our research has provided important new results for the seismic risk of Szeged and Debrecen, which show that it is necessary to heed the seismic safety in these cities.

Acknowledgements

This work have been carried out in the framework of the "CIPA-CT94-0238" project "Quantitative seismic zoning of the Circum-Pannonian region" and has been partly supported by the "AKP 96/2-447 2,5/40" project of the Hungarian Academy of Sciences. The authors extend gratitude to Professor G. Panza for numerous fruitful discussions and for his valuable remarks.

REFERENCES

AKI, K., *Strong motion seismology.* In *Strong Ground Motion Seismology, NATO ASI Series, Series C: Mathematical and Physical Sciences* (eds. Erdik, M. Ö., and Toksöz, M. N.) (D. Reidel Publishing Company, Dordrecht 1987) vol. 204, pp. 3–39.

AOUDIA, A., VACCARI, F., and SUHADOLC, P., *Seismic hazard assessment in the Tell Atlas, Algeria.* In *ESC Abstracts, XXV General Assembly, September 9–14* (Reykjavik, Iceland 1996) p. 84.

BARVITZ, A., LAKATOS, L., POGÁCSÁS, Gy., RUMPLER, J., SIMON, E., UJSZÁSZI, K., VAKARCS, G., VÁRKONYI, L., and VÁRNAI, P., 1990, *Magyarország tektonikai térképe (Tectonical Map of Hungary)* M = 1:800,000, Ph.D. thesis of Gy. Pogácsás.

BISZTRICSÁNY, E., *Mérnökszeismológia (Engineering Seismology)* (Akadémiai Kiadó 1974).

BONDÁR, I., BUS, Z., ŽIVČIĆ, M., COSTA, G., and LEVSHIN, A., *Rayleigh wave group and phase velocity measurements in the Pannonian basin.* In *Special Publications of the Geophysical Society of Greece, No. 6,* pages 73–86, 1996.

COSTA, G., PANZA, G. F., SUHADOLC, P., and VACCARI, F. (1993), *Zoning of the Italian Territory in Terms of Expected Peak Ground Acceleration Derived from Complete Synthetic Seismograms,* J. appl. geophys. *30,* 149–160.

CSÁK, B., HUNYADI, F., and VÉRTES, Gy. (1981), *Földrengések hatása az épitményekre (The Effects of Earthquakes on the Edifices),* Müszaki Könyvkiadó.

CSONTOS, L., NAGYMAROSY, A., HORVÁTH, F., and KOVAC, M. (1992), *Tertiary Evolution of the Intra-Carpıthian Area: A Model*, Tectonophysics *208*, 221–241.

EUROCODE 8 (1993), *Eurocode 8 Structures in Seismic Regions-Design-Part 1 General and Building*, Doc TC250/SC8/N57A.

FLORSCH, N., FÄH, D., SUHADOLC, P., and PANZA, G. F. (1991), *Complete Synthetic Seismograms for High-frequency Multimode Love Waves*, Pure appl. geophys. *136*, 529–560.

GERNER, P. (1995), *Catalogue of Earthquake Focal Mechanism Solutions for the Pannonian Region*, Geophysical Department of Eötvös University, Budapest.

GUSEV, A. A. (1983), *Descriptive Statistical Model of Earthquake Source Radiation and its Application to an Estimation of Short-period Strong Motion*, Geophys. J. R. Astron. Soc. *74*, 787–800.

GUTDEUTSCH, R., and ARIC, K., *Seismicity and neotectonics of the East Apline-Carpathian and Pannonian area*. In *The Pannonian Basin* (eds. Royden, L. H., and Horváth, F.) AAPG Memoir *45*, pp. 183–194, 1988.

HORVÁTH, F. (1993), *Towards a Mechanical Model of the Formation of the Pannonian Basin*, Techtonophysics *226*, 333–357.

HORVÁTH, F. (1995), *Phases of the Compression during the Evolution of the Pannonian Basin and its Bearing on Hydrocarbon Exploration*, Marine and Petroleum Geology *12*, 837–844.

KENNETT, B. L. N., and ENGDAHL, E. R. (1991), *Travel Times for Global Earthquake Location and Phase Identification*, Geophys. J. Int. *105*, 429–466.

KILÉNYI, E., and SEFARA, J., eds. *Pretertiary Basement Contour Map of the Carpathian Basin beneath Austria, Czechcoslovakia and Hungary*, Eötvös Loránd (Geophysical Institute of Hungary, Budapest 1989).

OROZOVA-STANISHKOVA, I. M., COSTA, G., VACCARI, F., and SUHADOLC, P. (1996), *Estimates of 1 Hz Maximum Acceleration in Bulgaria for Seismic Risk Reduction Purposes*, Tectonophysics *258*, 263–274.

PANZA, G. F. (1985), *Synthetic Seismograms: The Rayleigh Waves Modal Summation*, J. Geophys. Res. *58*, 125–145.

PANZA, G. F., VACCARI, F., COSTA, G., SUHADOLC, P., and FÄH, D. (1996), *Seismic Input Modelling for Zoning and Microzoning*, Earthquake Spectra *12*, 529–566.

POSGAY, K., ALBU, I., RÁNER, G., and VARGA, G., *Characteristics of the reflecting layers in the earth's crust and upper mantle in Hungary*. In *Reflection Seismology: A Global Perspective* (eds. BARAZANGI, M., and BROWN, L.) (Geodyn. Ser., AGU) (Washington, D.C. 1986) vol. 13, pp. 55–65.

RADULIAN, M., VACCARI, F., MANDRESCU, N., and PANZA, G. F. (2000), *Seismic Hazard of Romania: Deterministic Approach*, Pure appl. geophys. *157*, 221–247.

REISNER, G. I., and SHOLPO, V. N. (1975), *Distinguishing Zones of Seismic Risk According to Geological Data in the Territories of Czechoslovakia, Hungary and Roumania*, Veröff. Zentr. Inst. Phys. Erde *31*, 133–141.

STEGENA, L., GÉCZY, B., and HORVÁTH, F. (1975), *Late Cenozoic Evolution of the Pannonian Basin*, Tectonophysics *26*, 203–219.

SZABÓ, Z., and PÁNCSICS, Z., *A Pannon medence közetfizikai paraméterei (Parameters of the Rocks in the Pannonian Basin)* (Eötvös Loránd Geophysical Institute 1994).

SZEIDOVITZ, GY. (1986a), *Earthquakes in the Region of Komárom, Mór and Várpalota*, Geophys. Transact. *32*, 225–274.

SZEIDOVITZ, GY. (1986b), *The Dunaharaszti Earthquake January 12, 1956*, Acta Geod. Geophys. and Mont. *21*, 109–125.

SZEIDOVITZ, GY., and BUS, Z. (1995), *Seismological Investigations in the Kecskemét Area*, Acta Geod. Geophys. and Mont. *30*, 419–435.

TÓTH, L., MÓNUS, P., and ZSIROS, T. (1989), *The Berhida (Hungary) Earthquake of 1985*, Gerlands Beitr. Geophysik. *98*, 312–321.

VACCARI, F., COSTA, G., SUHADOLC, P., and PANZA, G. F. (1995), *Zonazione sismica deterministica al prim'ordine per l'area italiana*, Atti dei Convegni Lincei *122*, 117–126.

ZSIROS, T. (1985), *An Estimation of Seismic Hazard in Hungary*, Gerlands Beitr. Geophysik *94*, 111–122.

ZSIROS, T., and MÓNUS, P. (1984), *An Estimation of Maximum Ground Motions Caused by Earthquakes in Hungary*, Acta Geod. Geophys. and Mont. *19*, 433–449.

ZSIROS, T., MÓNUS, P., and TÓTH, L., *Hungarian Earthquake Catalog (456–1986)* (HAS Geodetical and Geophysical Research Institute 1988).

(Received April 25, 1998, revised December 7, 1998, accepted December 7, 1998)

To access this journal online:
http://www.birkhauser.ch

Pure appl. geophys. 157 (2000) 221–247
0033–4553/00/020221–27 $ 1.50 + 0.20/0

Pure and Applied Geophysics

Seismic Hazard of Romania: Deterministic Approach

M. Radulian,[1] F. Vaccari,[2,3] N. Mândrescu,[1] G. F. Panza[2,3] and
C. L. Moldoveanu[1]

Abstract—The seismic hazard of Romania is estimated in terms of peak-ground motion values—displacement, velocity, design ground acceleration (DGA)—computing complete synthetic seismograms, which are considered to be representative of the different seismogenic and structural zones of the country. The deterministic method addresses issues largely neglected in probabilistic hazard analysis, e.g., how crustal properties affect attenuation, since the ground motion parameters are not derived from overly simplified attenuation "functions," but rather from synthetic time histories. The synthesis of the hazard is divided into two parts, one that of shallow-focus earthquakes, and the other, that of intermediate-focus events of the Vrancea region.

The previous hazard maps of Romania completely ignore the seismic activity in the southeastern part of the country (due to the seismic source of Shabla zone). For the Vrancea intermediate-depth earthquakes, which control the seismic hazard level over most of the territory, the comparison of the numerical results with the historically-based intensity map show significant differences. They could be due to possible structural or source properties not captured by our modeling, or to differences in the distribution of damageable buildings over the territory (meaning that future earthquakes can be more spectacularly damaging in regions other than those regions experiencing damage in the past). Since the deterministic modeling is highly sensitive to the source and path effects, it can be used to improve the seismological parameters of the historical events.

Key words: Seismic hazard, deterministics modeling, Romania, Vrancea intermediate-depth earthquakes.

1. Introduction

The seismic hazard of Romania is relatively high, mainly due to the intermediate-depth earthquakes located in a confined focal volume at the Eastern Carpathians arc bend, in Vrancea region. One to five shocks with $M_w > 7$ occur here each century and are felt over a very large territory, from the Greek Islands to Scandinavia, and from Central Europe to Moscow. The largest and the most damaging subcrustal events reported from the beginning of the 19th century

[1] National Institute for Earth Physics, Bucharest, Romania, e-mail: mircea@infp.ro, e-mail: carmen@geosun0.univ.trieste.it, e-mail: mandrescu @infp.ro.

[2] The Abdus Salam International Centre for Theoretical Physics, SAND Group, Trieste, Italy.

[3] Dipartimento di Scienze della Terra, Universita' di Trieste, Italy, e-mail: vaccari@geosun0.univ.trieste.it, e-mail: panza@geosun0.univ.trieste.it.

occurred on October 26, 1802 ($M_w = 7.9$), November 26, 1829 ($M_w = 7.3$), January 11, 1838 ($M_w = 7.5$), November 10, 1940 ($M_w = 7.7$), and March 4, 1977 ($M_w = 7.4$). The moment magnitudes are from the catalog of ONCESCU *et al.* (1999). On the other hand, the crustal seismicity is moderate (M_{max}, for shallow events observed on the Romanian territory is 6.4) with only a few isolated active areas (Shabla, Făgăraş-Câmpulung, Banat, Crişana-Maramureş).

Several studies have been carried out to evaluate the seismic hazard of Romania, all based on probabilistic approaches (e.g., RADU and APOPEI, 1978; MÂN-DRESCU, 1990, 1984; LUNGU *et al.*, 1995). A totally different approach, developed by COSTA *et al.* (1992, 1993), is applied in the present paper to obtain a first-order deterministic mapping of the seismic hazard of Romania. The computations are performed using the modal summation method (PANZA, 1985; FLORSCH *et al.*, 1991) at regional scale, for one-dimensional average structural models and scaled

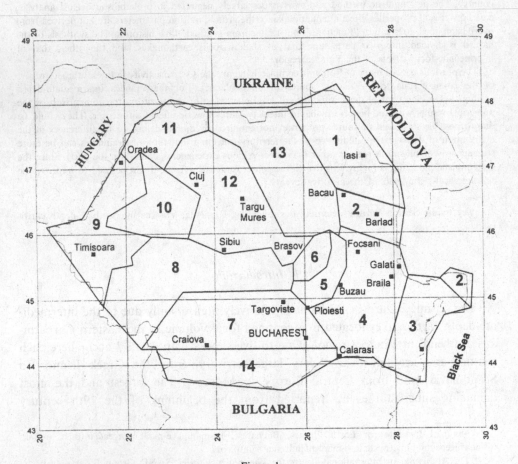

Figure 1

Map of Romania, showing the structural polygons. The boundaries are plotted as solid lines. The numbers of the structural units correspond with those given in Figure 2.

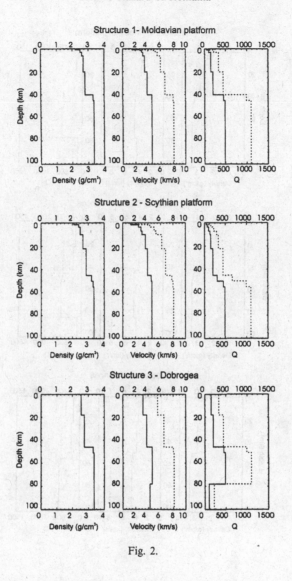

Fig. 2.

double-couple sources (GUSEV, 1983; RADULIAN *et al.*, 1998). The maximum ground acceleration, velocity and displacement in a given frequency range or any other parameter relevant to seismic engineering which is extracted from observed time series, can be estimated from simulated theoretical signals. This procedure also allows us to estimate the seismic hazard in those areas for which scarce (or no) historical or instrumental information is available, and to perform relevant parametric analyses. The peak values of the modeled ground motion, tested against the few available recorded values, are used for the estimation of the seismic hazard level.

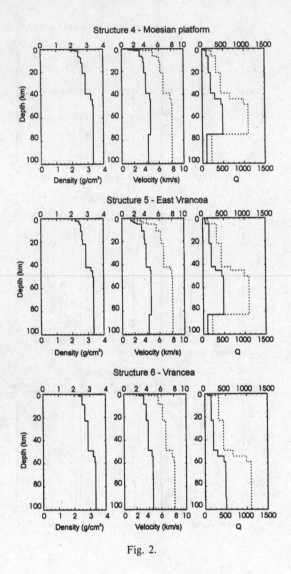

Fig. 2.

2. Input Data

Two kinds of input data are required by the computation algorithm: the structural and the source parameters. The Romanian territory is divided into regional polygons where average layered structures are specified (Fig. 1). The polygon boundaries, defined by RADULIAN *et al.* (this issue), roughly follow the contact between different tectonic units, or separate different lithospheres' properties. The structures at the political borders are correlated with data taken from OROZOVA-STANISHKOVA *et al.* (1996) for Bulgaria, and BUS *et al.* (this issue) for Hungary. The models (depth, density, *P*- and *S*-wave phase velocities and quality

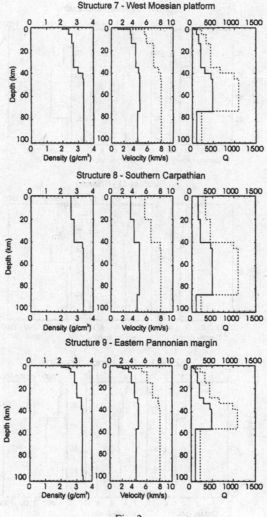

Fig. 2.

factor) are given in Figure 2. The average upper crust properties are defined on the basis of data from oil industry boreholes (GAVĂT, 1939; PARASCHIV, 1976, 1979). The depth of the Conrad and Moho discontinuities and the depth of the litho-sphere-astenosphere boundary are adopted considering geophysical (SOCOLESCU *et al.*, 1963, 1964, 1975; CONSTANTINESCU *et al.*, 1972; CORNEA *et al.*, 1981; LĂZĂRESCU *et al.*, 1983; ENESCU *et al.*, 1988; RĂDULESCU, 1988; RĂILEANU *et al.*, 1994), and seismological data (DEMETRESCU and ENESCU, 1960; IOSIF and IOSIF, 1973; ENESCU, 1987, 1992; ENESCU *et al.*, 1992).

For the structure below the lithosphere-asthenosphere boundary, a standard continental model (HARKRIDER, 1970) is adopted, with a low-velocity layer down to a depth of 200 km. The density is 3.35 g/cm³, and the parameters in the

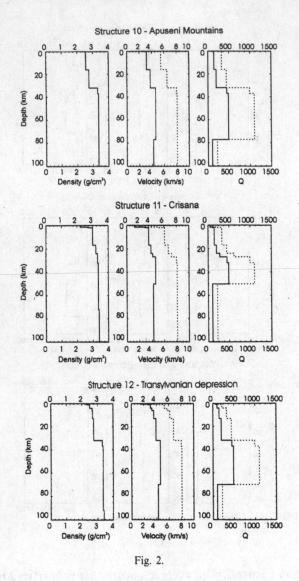

Fig. 2.

low-velocity channel are: P-wave velocity, $\alpha = 8.1$ km/s, S-wave velocity, $\beta = 4.25$ km/s, and quality factors, $Q_\alpha = 220$ and $Q_\beta = 100$. Structure number 6 corresponds to the Vrancea region where a highly confined focal volume, in the 60–200 km depth range, of subducted lithosphere is present. The low-velocity channel is, therefore, absent here and the parameters are: $\alpha = 8.1$ km/s, $\beta = 4.62$ km/s, $Q_\alpha = 1100$ and $Q_\beta = 690$.

The seismic sources are supposed to be distributed within the seismogenic zones defined by RADULIAN *et al.* (this issue), on the basis of geological, tectonic and seismicity information (Fig. 3). A representative focal mechanism is associated with each seismogenic zone. The scalar seismic moment associated with each source is

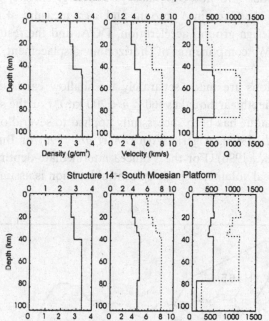

Figure 2

Average lithospheric structures for the regional polygons. The *P*- and *S*-wave velocities: in km/s, density: in g/cm³, and the corresponding quality factors (dotted line for *P* waves, solid line for *S* waves), are shown for the first 100 km of depth.

obtained by discretized and smoothed magnitude distribution, in accord with the procedure defined by COSTA *et al.* (1993). An updated version of the Romanian earthquake catalogue is considered by merging the catalogue of MUSSON (1996) for the Circum-Pannonian region and the recent catalogue of ONCESCU *et al.* (1999). The catalogue of fault plane solutions compiled by RADULIAN *et al.* (1997) for the Romanian earthquakes which occurred between 1929 and 1995 is used to extract the focal mechanism information. The characterization of the seismogenic zones is discussed in detail by RADULIAN *et al.* (this issue).

3. Seismic Hazard Computation

Starting from the structural models and seismic sources, *P-SV-* and *SH*-waves synthetic seismograms are computed on a dense grid (0.2° × 0.2°) by modal summation method (PANZA, 1985; FLORSCH *et al.*, 1991) for point double-couple sources. The computations are made for the frequency range from 0.005 to 1.0 Hz for which simple scaling of point-sources is still acceptable and detailed laterally heterogeneous models are not compulsory. For a given site, all the potentially dangerous sources are taken into account and, as a measure of the seismic hazard level, the largest design ground acceleration, DGA, and the resultant value among the N-S and E-W components of horizontal displacement and velocity are considered.

The computations are made separately for shallow earthquakes ($h < 60$ km), and intermediate-depth earthquakes ($60 \leq h < 200$ km). For the shallow events, this deterministic technique has been successfully applied to several other areas, such as Italy (COSTA *et al.*, 1993), Algeria (AOUDIA *et al.*, 1996) or Bulgaria (OROZOVA-STANISHKOVA *et al.*, 1996). For the Vrancea intermediate-depth events, due to the highly confined focal volume, a single epicentral location is assumed to be represen-

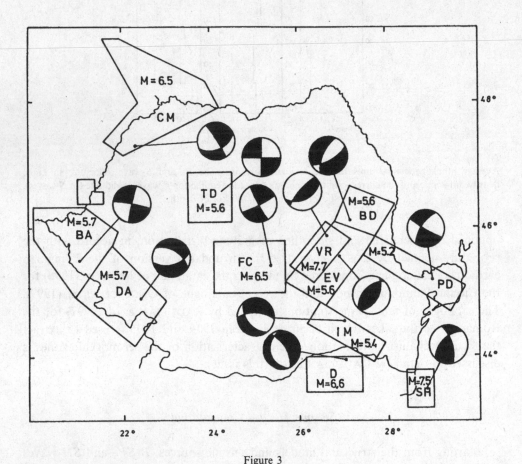

Figure 3

Seismogenic zones (solid contours): BD—Barlad Depression, PD—Predobrogean Depression, SH—Shabla, IM—Intramoesian Fault, D—Dobrogea, EV—East Vrancea, VR—Vrancea, FC—Făgăraş-Câmpulung, TD—Transylvanian Depression, DA—Danubian Depression, BA—Banat, CM—Crişana-Maramureş. For each seismogenic zone the maximum observed magnitude and typical fault plane solution are shown.

tative. To account for possible depth-dependent effects, two plausible focal depth values are considered: $h = 90$ km and $h = 150$ km.

The synthetic seismograms are scaled to the assumed scalar seismic moment: (a) for the shallow events, all with $h = 10$ km, we use GUSEV's (1983) empirical spectral laws, as reported by AKI (1987); (b) for the intermediate-depth earthquakes, we modified GUSEV's (1983) curves to account for the possible difference in the corner frequency between shallow and deep earthquakes (AKI and RICHARDS, 1980). For Romanian events with $M_w > 6$, the analysis of observed spectra indicates that, for a given magnitude, the corner frequencies of deep events are one order of magnitude higher than those of shallow earthquakes (RADULIAN et al., 1998). With this same scaling MOLDOVEANU and PANZA (1999) did model, for microzonation purposes, the seismic strong motion along a cross section, representative of the geological setting of Bucharest, due to the May 30, 1990, Vrancea event ($M_w = 6.9$).

The deterministic modeling can be extended to frequencies greater than 1 Hz by using the existing standard design response spectra (PANZA et al., 1996). The design ground acceleration (DGA) values are obtained by scaling the chosen normalized design response spectrum (normalized elastic acceleration spectra of the ground motion for 5% critical damping) with the response spectrum computed at frequencies below 1 Hz.

For the shallow earthquakes we use Eurocode 8 design response spectrum for class A soil (EC8, 1993).

For the intermediate-depth events we consider the results of LUNGU et al. (1995), who analyzed the strong ground motion records of the major Vrancea

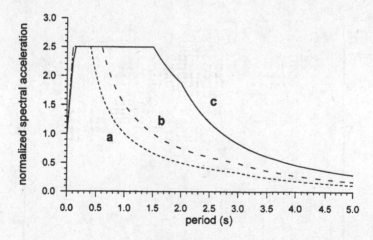

Figure 4

Comparative representation of different design response spectra used to construct the DGA maps for Romania: (a) Eurocode 8, soil A (EC 8, 1993)—dotted line; (b) Moldova region (LUNGU et al., 1995)—dashed line; (c) Bucharest city (LUNGU et al., 1995)—solid line. Spectra (b) and (c) are obtained from the Vrancea intermediate-depth earthquakes. The elastic response spectrum is normalized to the design ground acceleration.

earthquakes of March 1977 ($h = 94$ km, $M_w = 7.4$), August 30, 1986 ($h = 131$ km, $M_w = 7.1$), May 30, 1990 ($h = 89$ km, $M_w = 6.9$), and May 31, 1990 ($h = 79$ km, $M_w = 6.4$), and proposed two characteristic design response spectra, one, with regional validity, for the Moldova region, and the other, with local validity, for Bucharest city, shown in Figure 4.

4. Seismic Hazard Mapping

The seismic hazard, expressed in terms of peak ground displacement, velocity, and DGA values, corresponding to the shallow earthquakes, is shown in Figure 5. The largest values are obtained in the southeastern part, at the border with Bulgaria: 10 cm maximum displacement, 27 cm/s maximum velocity and 0.25 g DGA (where g is gravity acceleration), which imply a macroseismic intensity VIII, in the MSK-76 scale, as can be seen from Table 1 (MEDVEDEV, 1977). These values are due to the events in the Shabla zone (Bulgaria). The up-to-date Romanian

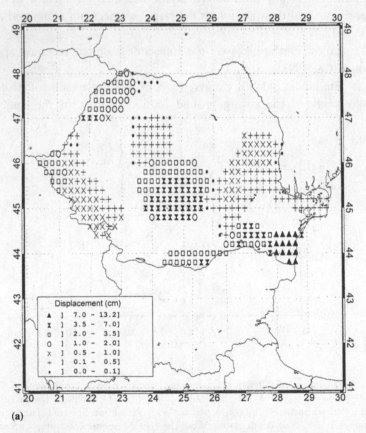

(a)

Fig. 5.

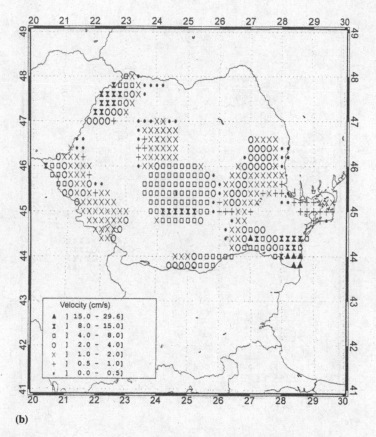

(b)

Fig. 5.

standard for seismic zoning (SR 11100, 1993), given in Figure 6, completely ignores
the seismic activity in the Shabla zone and consequently, in the southeast of
Romania, it indicates intensity VII, the intensity value which is due to the Vrancea
earthquakes. The seismic hazard is high all along the southern part of Dobrogea,
between the Black Sea shoreline and the bending to the north of the Danube river
(0.16 g close to Călărasi city). Relatively high values are obtained in the Câm-
pulung-Făgăraş region (0.09 g; 25°E, 45°N), Banat region (0.11 g; 20.4°E, 46.0°N),
and Crişana-Maramureş region (0.08 g; 22.0°E, 47.2°N), corresponding to intensity
VII. For the rest of the territory the DGA values are below 0.05 g.

   For the Vrancea strong subcrustal earthquakes two typical sources are analyzed:
(i) a source located in the upper part of the slab, at 90 km of depth with $M_w = 7.4$,
corresponding to the March 4, 1977 event, and (ii) a source located in the lower
part of the slab, at 150 km of depth with $M_w = 7.7$, corresponding to the November
10, 1940 event. The two focal depths selected in the computations are typical
average depths for the Vrancea earthquakes, and correspond to the maxima
observed in the depth distribution of the earthquake energy release (TRIFU and

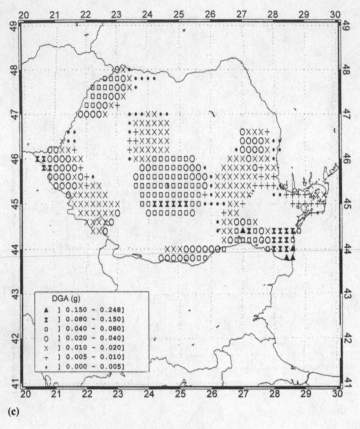

(c)

Figure 5

Seismic hazard map for shallow earthquakes. (a) Resultant displacement in cm for 1 Hz upper frequency; (b) resultant velocity in cm/s for 1 Hz upper frequency; (c) DGA, expressed in units of gravity acceleration (g).

Table 1

The intensity scale MSK-76 and associated average peak values of ground motion (MEDVEDEV, 1977)

Intensity (degree)	Acceleration (cm/s²)	Velocity (cm/s)	Displacement (cm)
V	25	2	1
VI	50	4	2
VII	100	8	4
VIII	200	16	8
IX	400	32	16
X	800	64	32

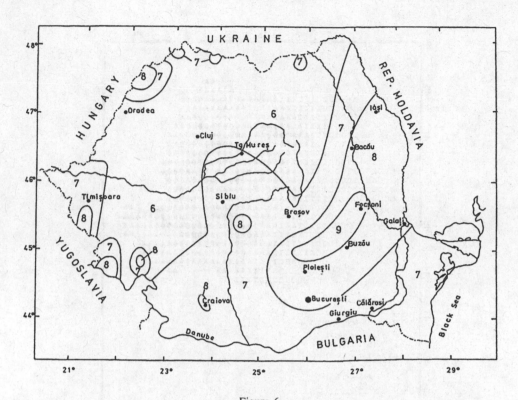

Figure 6
Standard seismic zoning of the Romanian territory (Romanian standard SR 11100/1, 1993). The intensity isolines (MSK scale) are represented by solid contours.

RADULIAN, 1991; ISMAIL-ZADEH *et al.*, this issue). The observed focal mechanism of the two events are close to each other, and we consider for both cases an average mechanism with the following parameters of the fault plane: 225° strike, 60° dip and 80° rake. The choice of a common fault plane solution is fully justified by the fact that 90% of the studied events, regardless of their magnitude, are characterized by a reverse faulting mechanism with the T-axis nearly vertical and the P-axis nearly horizontal (ENESCU, 1980; ENESCU and ZUGRĂVESCU, 1990; ONCESCU and TRIFU, 1987).

The seismic hazard map corresponding to source (i) is shown in Figure 7. The highest amplitudes are visible southeastward of the epicentral area. DGA values above 0.3 g are distributed over an area that extends to Galati to the east (0.39 g DGA, 95 cm/s maximum velocity and 32 cm maximum displacement at 28.0°E, 45.4°N) and to Târgoviste to the west (0.32 g, 45 cm/s maximum velocity and 23 cm maximum displacement at 25.6°E, 44.8°N). The maximum values are obtained in the area of Ploiesti, 50 km north of Bucharest (0.47 g DGA, 92 cm/s velocity and 33 cm displacement at 26.4°E, 44.8°N). In Bucharest the peak-ground motion parameters are: 0.23 g, 27 cm/s and 18 cm. From Table 1 we can see that these

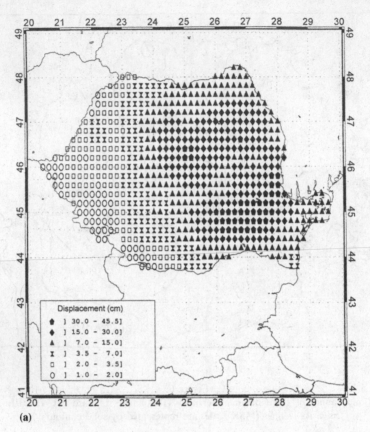

Fig. 7.

values correspond to an intensity VIII, the value observed in 1977 (MÂNDRESCU and RADULIÁN, 1999). High values are seen in the southeastern part of the Transylvanian basin: 0.34 g DGA, 69 cm/s velocity and 25 cm displacement at 25.4°E, 46.0°N. Values greater than 0.2 g are spread to the N-NE close to the border with the Republic of Moldova, to the S-SE to the latitude of 44.4°E, affecting most of the eastern sector of the Moesian platform (between 26°E and 28°E longitude), with the exception of Dobrogea, and to the NW in the Transylvanian depression reaching the Târgu Mures city (24.6°E, 46.6°N).

The seismic hazard map computed for source (ii) is shown in Figure 8. Generally, the distribution of the amplitude values is similar to the previous case. Two effects of the focal depth are however visible: the area of the near-epicenter local minimum is more developed, and the area of the largest values is shifted to larger epicentral distances (0.63 g DGA, 124 cm/s maximum velocity, 43 cm maximum displacement at 27.4°E, 44.4°N). The ground motion in the Bucharest area is particularly large: 0.52 g DGA, 105 cm/s maximum velocity,

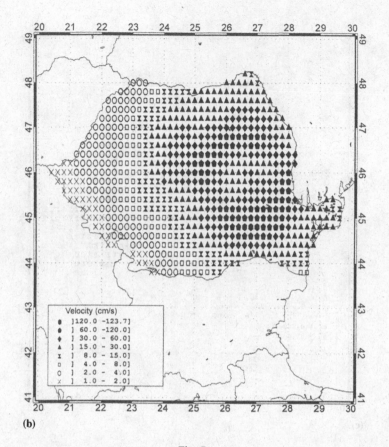

(b)

Fig. 7.

42 cm maximum displacement. DGA values greater than 0.3 g are seen over an extended area in the Moesian and Scythian platforms, oriented NE-SW. Similar values are obtained in the Transylvanian depression (the maximum value of 0.55 g DGA, 116 cm/s velocity and 40 cm displacement at 24.8°E, 46.8°N) and in the northern part of Moldova (0.58 g DGA, 114 cm/s velocity and 48 cm displacement at 26.2°E, 47.4°N). Significantly lower DGA values, in the range 0.25 g–0.06 g, are visible in an area situated to the NW of Vrancea.

The design response spectrum proposed by LUNGU et al. (1995) for Moldova region (Fig. 4) is adopted to compute the DGA values for the subcrustal earthquakes. To test the influence of the selected design response spectrum on the computed DGA values, considering source (ii), we use the design response spectrum proposed for Bucharest, which is specific for a soft soil. As we could expect, the DGA values obtained in this way (Fig. 9) are lower than the values shown in Figure 8.

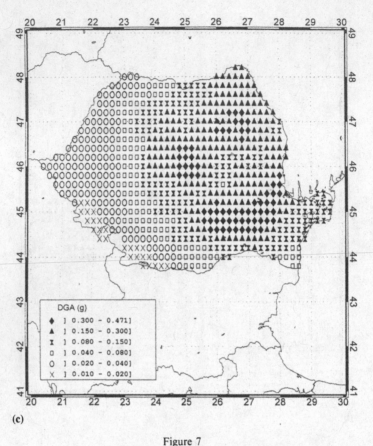

(c)

Figure 7

Seismic hazard map corresponding to source (i): (a) Resultant displacement in cm for 1 Hz upper frequency; (b) resultant velocity in cm/s for 1 Hz upper frequency; (c) DGA, expressed in units of gravity acceleration (g).

5. Synthetic Model against Observations

Strong motion data have been available in Romania since the 1977 major earthquake, when a SMAC-B accelerometer was operating in Bucharest. The maximum peak values (horizontal component) recorded in Bucharest in 1977 are: 15 cm, 65 cm/s and 194 cm/s², and the corresponding computed values are: 18 cm, 27 cm/s and 225 cm/s², respectively. Many accelerometers (SMA-1) were installed after 1977 and some of them triggered during the shocks of August 30, 1986, May 30 and May 31, 1990 (Table 2). For each of these events we have more than 10 digitized accelerograms available. We compute DGA, maximum synthetic velocity and displacement in correspondence of the sites of the strong motion stations, and we compare them with the values obtained from the records (Figs. 10–12).

For these events the computed DGA values are generally smaller than the observations, while velocities and displacements become more evenly distributed

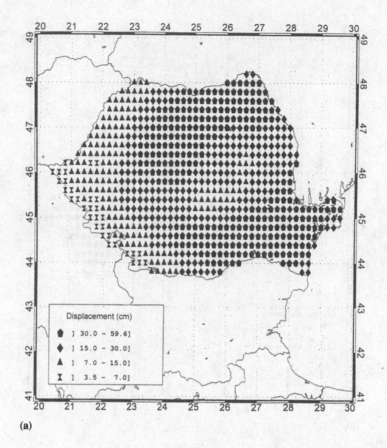

(a)

Fig. 8.

around the straight-line with slope 1. The large scatter in the data is compatible with local effects already reported in the literature. The systematic underestimation of the DGA can be reduced considering either smaller depths or larger magnitudes, as shown, for example for the event of May 30, 1990, in Figures 13 and 14.

There is substantial macroseismic information pertaining to Vrancea earthquakes. Reliable reports about earthquakes date back to the 14th century. The identification of the historical Vrancea intermediate-depth events is easier than that for the shallow earthquakes, since the area where intermediate-depth shocks are felt is extremely large. In all cases the maximum intensity is observed within a NE-SW elongated area, located southeastward of the epicenter. This area often includes several important cities, like Bucharest, Ploiesti, Buzău, Galati, Brăila, Focsani, Bârlad. Such a trend is well reproduced by our computations, especially for the 1977 case. When the focal depth increases, the maximum values are shifted SE (Fig. 8). In all cases, the computed maximum intensity occurs outside of the epicentral area, in agreement with the observations.

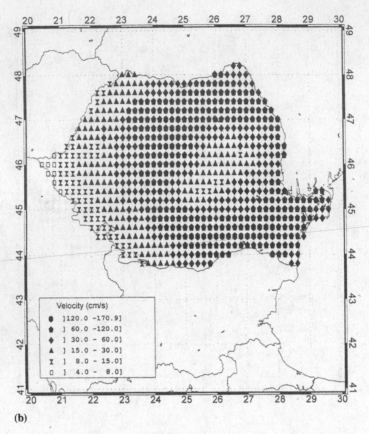

Fig. 8.

The deterministic approach creates, in the Transylvanian depression and north Moldova, intensity values that are at least one degree higher than the corresponding values proposed by the Romanian seismic zoning standard (SR 11100, 1993). The instrumental data are very scarce in these regions. It is therefore of crucial importance to install new strong-motion instruments and to intensify macroseismic investigation of historical events before any conclusion can be reached regarding the seismic hazard level here. In fact, the inspection of the available macroseismic data (macroseismic maps provided by the Institute for Geology and Geophysics, Kishinev) reveals the observation of intensity values in the Sibiu area similar with those recorded in Bucharest, Galati or Iasi, and in agreement with our computations.

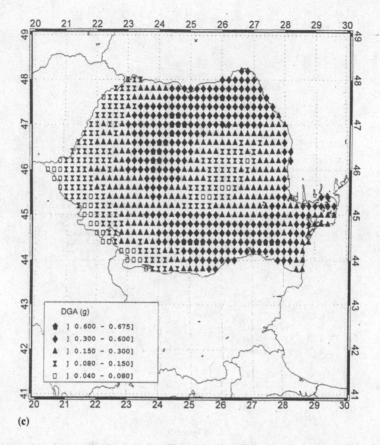

(c)

Figure 8

Seismic hazard map corresponding to source (ii): (a) Resultant displacement in cm for 1 Hz upper frequency; (b) resultant velocity in cm/s for 1 Hz upper frequency; (c) DGA, expressed in units of gravity acceleration (g).

6. Discussion and Conclusions

Using numerically simulated ground motion, a first-order deterministic evaluation of the seismic hazard of Romania is proposed. Several simplifying assumptions are adopted in the computation: simply scaled sources with prescribed average focal mechanism and depth for each seismogenic zone, and one-dimensional modeling of the structure. DGA is estimated by extrapolating the long period part of the response spectrum ($T \geq 1$ s), determined from the synthetic signal, with the design spectrum recommended by standard building codes.

The seismic hazard due to shallow earthquakes is generally moderate, with DGA values less than 0.1 g, with the exception of the southeastern part at the border with Bulgaria (Shabla zone). Other few limited areas are characterized by DGA values around 0.1 g: Făgăraş-Câmpulung zone in the Southern Carpathians,

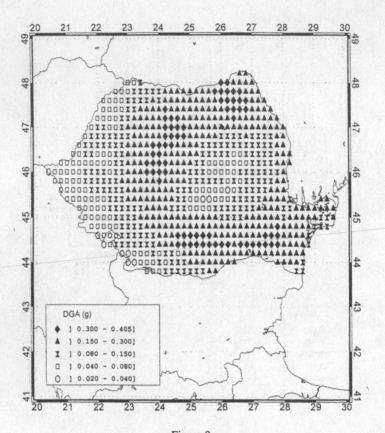

Figure 9

Seismic hazard map (DGA values) corresponding to source (ii) when the design response spectrum of Bucharest is used.

and Banat and Crişana-Maramureş zones at the eastern margin of the Pannonian basin.

The seismic hazard level of the Romanian territory is mostly controlled by the Vrancea intermediate-depth seismicity. We consider two typical cases associated

Table 2

Source parameters of the Vrancea major earthquakes, in this century

Date	Time	Lat.[1] (°N)	Lon.[1] (°E)	Depth[1] (km)	Depth[2] (km)	M_w^2	Strike	Dip	Rake
1940 11 10	01:39	45.8	26.7	150	150	7.7	225	60	80
1977 03 04	19:21	45.8	26.8	94	90	7.4	225	60	80
1986 08 30	21:28	45.5	26.5	131	130	7.1	227	65	104
1990 05 30	10:40	45.9	26.9	89	60	6.9	239	63	101
					90	7.4			
1990 05 31	00:17	45.8	26.9	79	79	6.4	310	70	90

Notes: [1] Oncescu and Bonjer (1997); [2] Values used in the computations of the synthetic seismograms.

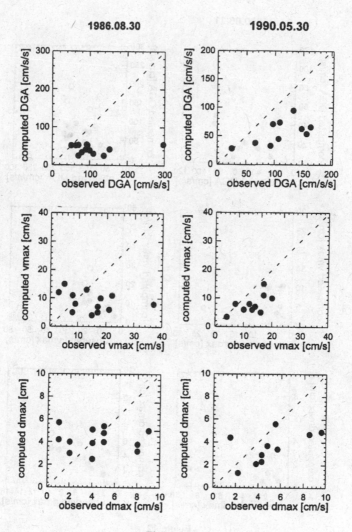

Figure 10

Computed DGA, maximum velocity and maximum displacement values versus corresponding observed values for the august 30, 1986, Vrancea earthquake ($h = 130$ km, $M_w = 7.1$). The velocity and displacement are integrated from the available accelerograms. The line of unit slope is drawn as reference.

Figure 11

Computed DGA, maximum velocity and maximum displacement values versus corresponding observed values for the May 30, 1990, Vrancea earthquake ($h = 89$ km, $M_w = 6.9$). The velocity and displacement are integrated from the available accelerograms. The line of unit slope is drawn as reference.

with the largest and most damaging earthquakes recorded this century: November 10, 1940 ($M_w = 7.7$) and March 4, 1977 ($M_w = 7.4$). DGA values greater than 0.3 g are obtained over an extended surface. The distribution of the peak values numerically determined correlates well with the values recorded in the area situated eastward and southward of the Carpathians arc. As our results show, it is of

1990.05.31

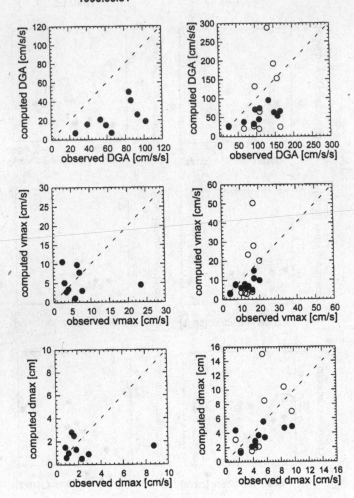

Figure 12

Computed DGA, maximum velocity and maximum displacement values versus corresponding observed values for the May 31, 1990, Vrancea earthquake ($h = 79$ km, $M_w = 6.4$). The velocity and displacement are integrated from the available accelerograms. The line of unit slope is drawn as reference.

Figure 13

Computed DGA, maximum velocity and maximum displacement versus corresponding observed values for the May 30, 1990 Vrancea earthquake, $M_w = 6.9$, $h = 90$ km (●), $h = 60$ km (○). The velocity and displacement are integrated from the available accelerograms. The line of unit slope is shown for reference. The depth of 60 km, consistent with the upper bound of the CMT hypocentral determination, improves the modeling of the observed ground motion field, and it is consistent with the waveform modeling made by MOLDOVEANU and PANZA (1999).

extreme importance to install new strong motion instruments in Transylvania to constrain the seismic hazard level in this area since, according to our modeling, the ground motion values due to the Vrancea earthquakes (especially for the deeper events) could be higher than believed, on the basis of very poor observations.

Therefore in SE Romania and Transylvania the seismic hazard may be higher by one degree of MSK intensity than predicted by the up-to-date Romanian standard for seismic zoning (SR 11100, 1993), which is obviously requiring a significant revision.

Further investigations are necessary to establish if the strong asymmetrical distribution of the motion amplitude on the NW-SE direction, across the Carpathians arc, is real. For this purpose two lines of research must be pursued: (a) systematic analysis of macroseismic data, including historical documents; (b) introduction of sources with finite dimensions, to account for possible effects on the radiation pattern.

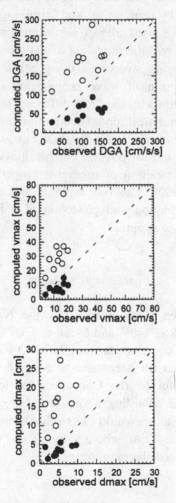

Figure 14

Computed DGA, maximum velocity and maximum displacement versus corresponding observed values for the May 30, 1990 Vrancea earthquake, $h = 90$ km, $M_w = 6.9$ (●), $M_w = 7.4$ (○). The observed velocity and displacement are integrated from the available accelerograms. The line of unit slope is drawn as reference.

The comparison with the instrumentally recorded accelerograms reveals that satisfactory predictions of the ground motion can be made, even allowing a few degrees of freedom for the source and medium. To take into account the uncertainties associated with the input parameters, extensive parametric analysis is required. Possible scenarios can be easily constructed for different source and structural parameters, which can be subsequently considered by civil engineers in the design of new seismo-resistant constructions and in the reinforcement of the existing built environment. As our study confirms, the seismic hazard level is very sensitive to the variation of the maximum magnitude, average depth and focal mechanism of the source. Therefore considerable attention must be paid to the possibility of properly defining these parameters by geophysical and seismological constrains.

The technique we have used allows significant improvements that may be required by an even more realistic modeling of the ground motion. It can, in fact, deal with (i) laterally varying structures, using either the hybrid method which combines, for the description of wave propagation in anelastic heterogeneous media, the modal summation with the finite difference techniques (FÄH *et al.*, 1990; FÄH, 1992), or the fully analytical modal summation method, extended to laterally varying media (VACCARI *et al.*, 1989; ROMANELLI *et al.*, 1996, 1997); and (ii) extended sources that account for possible complexities of the rupture process, and are represented as a superposition of subevents properly weighted and shifted in time (VACCARI *et al.*, 1990). The full exploitation of these possibilities of the method requires the production and collection of a large amount of data and evidence that is presently in progress.

Acknowledgements

This research has been made possible by the NATO Linkage Grant EN-VIR.LG. 960916, by MURST (40% and 60%), by COPERNICUS Project, ERBCI-PACT 94-0238, and is a contribution to UNESCO-IGCP Project 414 "Realistic Modelling of Seismic Input for Megacities and Large Urban Areas." One of the Authors (CLM) is grateful to the Consorzio per lo Sviluppo Internazionale, Universita' di Trieste, for awarding a one-year scholarship at Dipartimento di Scienze della Terra. The authors would like to thank two anonymous reviewers and the associated editor, Cezar Trifu, whose valuable and constructive comments and suggestions greatly enhanced the manuscript.

Realistic Modelling of Seismic Input for Megacities and Large Urban Areas (project 414)

REFERENCES

AKI, K., *Strong motion seismology*. In *Strong Ground Motion Seismology* (eds. Erdik, M. Ö., and Toksöz, M. N.) NATO ASI Series C: Mathematical and Physical Sciences (D. Reidel Publishing Company, Dordrecht 1987) *204*, 3–39.

AKI, K., and RICHARDS, P. (1980) *Quantitative Seismology, Theory and Methods* (W.H. Freeman and Company, San Francisco 1980).

AOUDIA, A., VACCARI, F., and SUHADOLC, P. (1996) *Seismic hazard assessment in the Tell Atlas, Algeria*. In *ESC Abstracts*, XXV Gen. Ass., Reykjavik, Iceland, September 9–14, 1996.

BUS, Z., SZEIDOVITZ, G., and VACCARI, F. (this issue), *Synthetic Seismogram Based Deterministic Seismic Zoning for the Hungarian Part of the Pannonian Basin*, Pure appl. geophys., this issue.

CONSTANTINESCU, L., CORNEA, I., and ENESCU, D. (1972), *Structure de la Croûte Terrestre de Roumanie d'après les Données Géophysiques*, Rev. Roum. Géol. Géophys., Géogr., Géophys. *16*, 3–20.

CORNEA, I., RĂDULESCU, F., POMPILIAN, A., and SOVA, A. (1981), *Deep Seismic Sounding in Romania*, Pure appl. geophys. *119*, 1144–1156.

COSTA, G., PANZA, G. F., SUHADOLC, P., and VACCARI, F. (1992) *Zoning of the Italian Region with Synthetic Seismograms Computed with Known Structural and Source Information*, Proc. 10-th WCEE, July 1992, Madrid, Balkema, 435–438.

COSTA, G., PANZA, G. F., SUHADOLC, P., and VACCARI, F. (1993), *Zoning of the Italian Territory in Terms of Expected Peak Ground Acceleration Derived from Complete Synthetic Seismograms*, J. Appl. Geophys. *30*, 149–160.

DEMETRESCU, C., and ENESCU, D. (1960), *Contributions to the Knowledge of the Earth Crust Structure in Romania*, St. Cerc. Astr., Seismologie *1*, 11–16 (in Romanian).

EC 8 (1993), *Eurocode 8 Structures in Seismic Regions—Design—Part 1: General and Building*, Doc TC250/SC8/N57A.

ENESCU, D. (1980), *Contributions to the Knowledge of the Focal Mechanism of the Vrancea Strong Earthquake of March 4, 1977*, Rev. Roum. Géol., Géophys., Géogr., Ser., Géophys. *24*, 3–18.

ENESCU, D. (1987), *Contributions to the Knowledge of the Lithosphere Structure in Romania on the Basis of Seismic Data*, St. Cerc. Geol. Geofiz. Geogr., Ser. Geofizica *25*, 20–27.

ENESCU, D., and ZUGRĂVESCU, D. (1990), *Geodynamical Considerations Regarding the Eastern Carpathians Arc Bend, Based on Studies on Vrancea Earthquakes*, Rev. Roum. Géophys. *34*, 17–34.

ENESCU, D. (1992), *Lithosphere Structure in Romania. I. Lithosphere Thickness and Average Velocities of Seismic Waves P and S. Comparison with other Geophysical Data*, Rev. Roum. Phys. *37*, 623–639.

ENESCU, D., POMPILIAN, A., and BĂLĂ, A. (1988), *Distribution of the Seismic Wave Velocities in the Lithosphere of Some Regions of Romania*, Rev. Roum. Géol. Géophys., Géogr., Géophys. *32*, 3–11.

ENESCU, D., DANCHIV, D., and BĂLĂ, A. (1992), *Lithosphere Structure in Romania. II. Thickness of Earth Crust. Depth-dependent Propagation Velocity Survey for P and S Waves*, St. Cerc. Geol. Geofiz. Geogr., Ser. Geofizica *30*, 3–19.

FÄH, D., SUHADOLC, P., and PANZA, G. F. (1990), *Estimation of Strong Ground Motion in Laterally Heterogeneous Media: Modal Summation—Finite Differences*, Proceedings of the 9th European Conference of Earthquake Engineering, Sept. 11–16, 1990, Moscow *4A*, 100–109.

FÄH, D. (1992), *A Hybrid Technique for the Estimation of the Strong Ground Motion in Sedimentary Basins*. Ph.D. Thesis Nr. 9767, Swiss Federal Institute of Technology, Zürich.

FLORSCH, N., FÄH, D., SUHADOLC, P., and PANZA, G. F. (1991), *Complete Synthetic Seismograms for High-frequency Multimode Love Waves*, Pure appl. geophys. *136*, 529–560.

GAVĂT, I. (1939), *Sur le anomalies du gradient horizontal de "G" aux confins des Subcarpates et de la Plaine Roumaine au point de vue de la prospection de pétrole*, Mon. Pétrol Roum. *35*, 1–14, Bucharest.

GUSEV, A. A. (1983), *Descriptive Statistical Model of Earthquake Source Radiation and its Applicability to an Estimation of Short-period Strong Motion*, Geophys. J. R. Astron. Soc. *74*, 787–800.

IOSIF, T., and IOSIF, S. (1973), *Data on the Crust and Upper Mantle*, St. Cerc. Geol. Geofiz. Geogr., Ser. Geofizica *11*, 203–219 (in Romanian).

HARKRIDER, D. (1970), *Surface Waves in Multilayered Elastic Media. Part II. Higher Mode Spectra and Spectral Ratios from Points Sources in Plane Layered Earth Models*, Bull. Seismol. Soc. Am. *60*, 1937–1987.

ISMAIL-ZADEH, A. T., PANZA, G. F., and NAIMARK, B. M. (this issue), *Stress in the Descending Relic Slab beneath the Vrancea Region, Romania*, Pure appl. geophys. (this issue).

LĂZĂRESCU, V., CORNEA, I., RĂDULESCU, F., and POPESCU, M. (1983), *Moho Surface and Recent Crustal Movements in Romania: Geodynamic Connections*, An. Inst. Geol. Geofiz. *63*, 163–168.

LUNGU, D., CORNEA, T., CRAIFĂLEANU, I., and ALDEA, A. (1995), *Seismic Zonation of Romania Based on Uniform Hazard Response*, Proc. Fifth International Conference on Seismic Zonation, October 17–19, 1995, Nice, France.

MÂNDRESCU, N. (1984), *Geological Hazard Evaluation of Romania*, Engineering Geology *20*, 39–47.

MÂNDRESCU, N. (1990), *Data Concerning Seismic Risk Evaluation*, Natural Hazards *3*, 249–259.

MÂNDRESCU, N., and RADULIAN, M., *Seismic microzoning of Bucharest (Romania): A critical review*. In *Vrancea Earthquakes: Tectonics, Hazard and Risk Mitigation* (eds. Wenzel, F., Lungu, D., and Novak, O.) (Kluwer Academic Publishers, 1999) pp. 163–174.

MEDVEDEV, S. V. (1977), *Seismic Intensity Scale MSK-76*, Publ. Inst. Geophys. Pol. Acad. Sc. *117*, 95–102.

MOLDOVEANU, C. L., and PANZA, G. F., *Modelling for microzonation purposes, of the seismic ground motion in Bucharest, due to the Vrancea earthquake of May 30, 1990*. In *Vrancea Earthquakes: Tectonics, Hazard, and Risk Mitigation* (eds. Wenzel, F., Lungu, D., and Novak, O.) (Kluwer Academic Publishers 1999) pp. 85 98.

MUSSON, R. M. W., *An earthquake catalogue for the Circum-Pannonian Basin*. In *Seismicity of the Carpatho-Balcan Region*, Proc. of XVth Congress of the Carpatho-Balcan Geol. Ass. (ed. Papanikolaou, D., and Papoulia, J.) (Athens 1996) pp. 233–238.

ONCESCU, M. C., and TRIFU, C.-I. (1987), *Depth Variation of the Moment Tensor Principal Axes in Vrancea (Romania) Seismic Region*, Ann. Geophisicae *5B*, 149–154.

ONCESCU, M. C., and BONJER, K.-P. (1997), *A Note on the Depth Recurrence and Strain Release of Large Vrancea Earthquakes*, Tectonophysics *272*, 291–302.

ONCESCU, M. C., MÂRZA, V. I., RIZESCU, M., and POPA, M. (1999), *The Romanian earthquake catalogue between 984–1996*. In *Vrancea Earthquakes: Tectonics, Hazard and Risk Mitigation* (eds. Wenzel, F., Lungu, D., and Novak, O.) (Kluwer Academic Publishers 1999) pp. 43–49.

OROZOVA-STANISHKOVA, I. M., COSTA, G., VACCARI, F., and SUHADOLC, P. (1996), *Estimates of 1 Hz Maximum Acceleration in Bulgaria for Seismic Risk Reduction Purposes*, Tectonophysics *258*, 263–274.

PANZA, G. F. (1985), *Synthetic Seismograms: The Rayleigh Waves Modal Summation*, J. Geophys. *58*, 125–145.

PANZA, G. F., VACCARI, F., COSTA, G., SUHADOLC, P., and FÄH, D. (1996), *Seismic Input Modelling for Zoning and Microzoning*, Earthquake Spectra *12*, 529–566.

PARASCHIV, D. (1976), *The Contribution of the Paleorelief to the Hydrocarbons Deposits Formation in Romania*, Rev. Roum. Géol. Géophys., Géogr., Géogr. *20*, 81–88.

PARASCHIV, D., (1979) *The Moesian Platform and its Hydrocarbon Deposits* (Academy Publishing House, Bucharest 1979) (in Romanian).

RADU, C., and APOPEI, I., (1978) *Macroseismic field of the Romanian earthquakes*, Proc. of the Symp. on the *Analysis of Seismicity and on Seismic Risk* (eds. Karnik, V., and Schenkova, Z.), October 17–22, 1977 (Liblice, Prague 1978).

RĂDULESCU, F. (1988), *Seismic Models of the Crustal Structure in Romania*, Rev. Roum. Géol. Géophys., Géogr., Géophys. *32*, 13–17.

RADULIAN, M., MÂNDRESCU, N., POPESCU, E., UTALE, A., and PANZA, G. F. (1997), *Seismicity and Stress Field Characteristics for the Seismogenic Zones of Romania*, EEC Technical Report, Project CIPA CT94-0238.

RADULIAN, M., MÂNDRESCU, N., VACCARI, F., and PANZA, G. F. (1998), *Deterministic Seismic Hazard Assessment of Romania*, EGS Abstracts XXIII Gen. Ass., Nice, 20–24 April, 1998.

RADULIAN, M., MÂNDRESCU, N., POPESCU, E., UTALE, A., and PANZA, G. F. (2000), *Characterization of the Romanian Seismogenic Zones*, Pure appl. geophys. *157*, 57–77.

RĂILEANU, V., DIACONESCU, C., and RADULESCU, F. (1994), *Characteristics of Romanian Lithosphere from Deep Seismic Reflection Profiling*, Tectonophysics *139*, 165–185.

ROMANELLI, F., BING, Z., VACCARI, F., and PANZA, G. F. (1996), *Analytical Computation of Reflexion and Transmission Coupling Coefficients for Love Waves*, Geophys. J. Int. *125*, 132–138.

ROMANELLI, F., BEKKEVOLD, J., and PANZA, G. F. (1997), *Analytical Computation of Coupling Coefficient in Non-Poissonian Media*, Geophys. J. Int. *125*, 205–208.

SOCOLESCU, M., POPOVICI, D., and VISARION, M. (1963), *Mohorovicic Surface in Eastern Carpathians and Transylvanian Basin, Resulted from Gravimetric Data*, St. Cerc. Geofiz. *1*, 1–10 (in Romanian).

SOCOLESCU, M., POPOVICI, D., VISARION, M., and ROȘCA, V. (1964), *Structure of the Earth's Crust in Romania as Based on the Gravimetric Data*, Rev. Roum. Géol. Géophys., Géogr., Géophys. *8*, 3–11.

SOCOLESCU, M., AIRINEI, S., CIOCÁRDEL, R., and POPESCU, M. (1975), *Physics and Structure of the Crust in Romania* (Technical Press, Bucharest 1975) (in Romanian).

SR 11100 (1993), *Seismic Zoning. Macrozoning of the Territory of Romania*, Institutul Român de Standardizare (IRS, in Romanian).

TRIFU, C.-I., and RADULIAN, M. (1991), *Frequency Magnitude Distribution of Earthquakes in Vrancea: Relevance for a Discrete Model*, J. Geophys. Res. *96*, 4301–4311.

VACCARI, F., GREGERSEN, S., FURLAN, M., and PANZA, G. F. (1989), *Synthetic Seismograms in Laterally Heterogeneous, Anelastic Media by Modal Summation of P-SV Waves*, Geophys. J. Int. *99*, 285–295.

VACCARI, F., SUHADOLC, P., and PANZA, G. F. (1990), *Irpinia, Italy, 1980 Earthquake: Waveform Modelling of Strong Motion Data*, Geophys. J. Int. *101*, 631–647.

(Received June 30, 1998, revised December 10, 1998, accepted May 31, 1999)

To access this journal online:
http://www.birkhauser.ch

Pure appl. geophys. 157 (2000) 249–267
0033–4553/00/020249–19 $ 1.50 + 0.20/0

Pure and Applied Geophysics

Estimation of Site Effects in Bucharest Caused by the May 30–31, 1990, Vrancea Seismic Events

C. L. Moldoveanu,[1,2] Gh. Marmureanu,[1] G. F. Panza[2,3] and F. Vaccari[2,4]

Abstract—The Vrancea region seismicity, characterized by focal depths larger than 60 km and major events with magnitudes $M_w \geq 6.9$, is responsible for the most destructive effects experienced in the Romanian territory, and may seriously affect high risk construction located on a wide area, from Central to Eastern Europe. This seismogenic volume must be taken into account both for seismic hazard analysis at the regional level (southeastern Europe) and national level (Romania and Bulgaria) as well as for microzonation studies of the highly populated cities located in the range of influence of this source. Since about four destructive earthquakes occur every century in Vrancea, the microzonation of Bucharest, the main city exposed to the potential damages due to these strong intermediate-depth shocks, represents an essential step towards the mitigation of the local seismic risk. Two main approaches can be considered for the evaluation of the local seismic hazard: (a) collection and extended use, for engineering purposes, of the recorded strong motion data, and (b) advanced modelling techniques that allow us the computation of a realistic seismic input, which can compensate for the lack of strong motion records, actually available only for a few events that occurred in the last 20–30 years.

Using a ground motion simulation technique that combines modal summation and finite differences, we analyze, along a geologic profile representative of the Bucharest area, the differences in the expected ground motion when two source mechanisms corresponding to the May 30–31, 1990, intermediate-depth Vrancea earthquakes, typical events for the Vrancea seismogenic zone, are considered. All three components of motion are influenced by the presence of the deep alluvial sediments, the strongest local effect being visible in the transversal (T) one, both observed and computed. The details of the local effects vary with varying the earthquake scenario, R and V components being very sensitive. Therefore, for a reliable determination of the seismic input all three components of motion (R, V and T) should be used.

Key words: Bucharest, Vrancea earthquakes, strong ground motion, 2-D numerical modelling, variability of local soil effects.

[1] National Institute for Earth Physics, P.O. Box MG 2, 7600 Bucharest, Romania. e-mail: carmen@geosun0.univ.trieste.it. e-mail: marmur@infp.infp.ro

[2] Dipartimento di Scienze della Terra, Universita' degli Studi di Trieste, Via E. Weiss 4, 31427 Trieste, Italy. e-mail: panza@geosun0.univ.trieste.it. e-mail: vaccari@geosun0.univ.trieste.it

[3] The Abdus Salam International Center for Theoretical Physics, SAND Group, Trieste, Italy.

[4] CNR-Gruppo Nazionale per la Difesa dia Terremoti, Roma, Italy.

1. Introduction

Bucharest, the capital of Romania, is heavily affected by the Vrancea strong intermediate-depth events, and the presence of more than two million inhabitants, together with a remarkable number of high seismic risk vulnerable buildings and infrastructures, makes the microzonation of the city a goal of main importance to achieve. During this century, the major events ($M_w \geq 6.5$) originating in Vrancea occurred in: 1904—$M_w = 6.6$, 1908—$M_w = 7.1$, 1912—$M_w = 6.7$, 1940—$M_w = 6.5$, 1940—$M_w = 7.7$, 1945—$M_w = 6.8$, 1945—$M_w = 6.5$, 1977—$M_w = 7.4$, 1986— $M_w = 7.1$, 1990—$M_w = 6.9$ (ONCESCU *et al.*, 1999). These epicenters are shown in

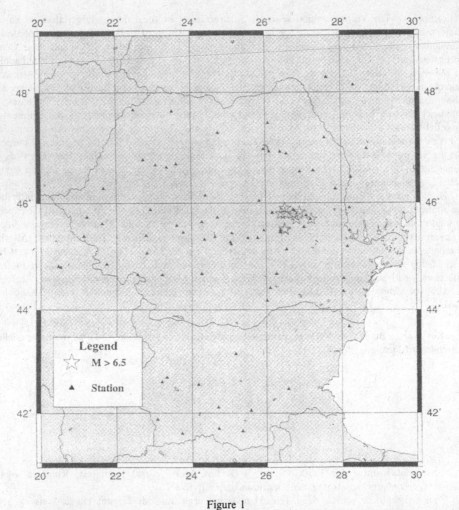

Figure 1

Location of the intermediate-depth events with $M_w \geq 6.5$ (☆) that occurred in Vrancea during this century. Triangles represent the seismic stations presently used by the National Institute for Earth Physics, Bucharest (NIEP) for locating the earthquakes.

Figure 1 together with the stations currently used by the National Institute for Earth Physics, Bucharest, for the events' location.

The evaluation of the seismic input at a given site is of key importance for engineering purposes and it can be performed following one of the two main approaches: (1) the collection and processing of the strong motion records obtained by means of a dense seismic network, and (2) the use of advanced modelling techniques for the computation of realistic seismic ground motion. The ideal situation is represented by the possibility of following both ways and calibrating the modelling with the available recordings. In practice, strong motion data are very scarce and correspond only to events that occurred in the last 20–30 years.

Exploiting the accumulated information concerning seismic sources, sampled medium and local soil conditions, together with realistic ground-motion simulation techniques, it is now possible to estimate, for microzonation purposes, the local behavior of a given site. Whenever possible, the complementary use of the two approaches should be followed because of (1) the high installation and operation cost of a dense permanent seismic network, and (2) the necessity to calibrate with observations the synthetic signals obtained using the geological geotechnical knowledge accumulated for the investigated region.

The Bucharest area represents a typical case in which the complementary use of modelling and data processing may allow us to obtain quite useful predictions of the expected ground motion, since only a few strong motion records of the last three strong Vrancea events are available. For microzonation purposes, MOLDOVEANU and PANZA (1999), making use of a simplified geotechnical profile, both for the regional and local structures, compute the seismic ground motion (SH and P-SV waves) along a representative profile in Bucharest indicated on the city sketch in Figure 2. As scenario earthquake they considered the strong Vrancea event of May 30, 1990 ($M_w = 6.9$). The description of the seismic wavefield generated by a given seismic source in a complex geological structure is performed with a hybrid method that successfully combines the analytical technique of modal summation (PANZA, 1985, 1993; VACCARI et al., 1989; FLORSCH et al., 1991; ROMANELLI et al., 1996), and the numerical technique of finite differences (FÄH, 1991; FÄH and PANZA, 1994; FÄH et al., 1994). Although no data fitting is made and relatively simple source and structural models are considered, the computed accelerograms are in good agreement with the only available records in Bucharest from Magurele station (44.347°N, 26.030°E), which is located on the local profile considered, in the southern part of the city (indicated in Fig. 2), and with the observed local site effects in Bucharest. In the present paper we analyze the variation of the local site effects when changing the scenario earthquake from May 30, 1990, to May 31, 1990, Vrancea events. The two focal mechanisms considered in this study are representative for the intermediate-depth seismicity of the region and, therefore, the results of our simulations allow a realistic estimation of the possible strong motion behavior in Bucharest induced by the Vrancea source.

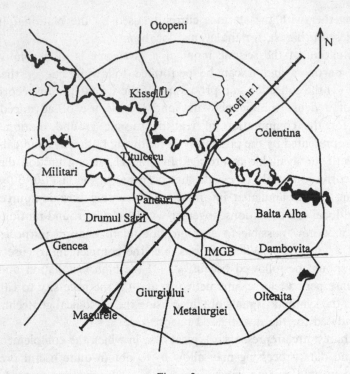

Figure 2
Bucharest city sketch and the cross section considered for computations. Bore-holes (ticks) along the profile (thick line); the position of Magurele station is indicated by a full triangle.

2. Brief Description of the Vrancea Region

The Vrancea region, localized in the rectangle delimited by latitude 45°–46°N and longitude 26°–27°E, beneath the bending of Eastern Carpathian Arc, is characterized by a very well confined and persistent subcrustal seismic activity. In this seismogenic source originate about 10–15 events per month ($2.5 < M_L < 5.5$), and three to five strong events ($M_w \sim 7.0$) per century. The volume in which the intermediate-depth earthquakes occur is a parallelepiped about 100 km long, 40 km wide, with a vertical extension from 50–60 km to 160–170 km depth. The subcrustal seismic activity concentrates within an epicentral area of about 3000 km², NE–SW oriented, that partly overlaps the epicentral area of the crustal events (RADULIAN *et al.*, 1996a; this issue). The location of both intermediate-depth and shallow events reported in the Vrancea region catalog (MOLDOVEANU *et al.*, 1998) for the period from 1932.1.1 to 1988.8.1 is presented in Figure 3. In the depth range from about 40 km to about 60 km, a gap is observed between the crustal and subcrustal seismic activity. The five major earthquakes ($M_0 \geq 10^{19}$ N·m, $M_w \geq 6.9$) that occurred in Vrancea (in 1980, 1940, 1977, 1986 and 1990) during this century (ONCESCU *et al.*, this issue) caused extensive damage and many casualties, not only in Romania, but also in other parts of Europe.

Several models have been proposed to explain the main aspects of the tectonic processes in Vrancea, and they are briefly summarized in the following.

Accordingly with MCKENZIE (1970, 1972), the subcrustal seismicity occurs in a vertical relic slab sinking into the asthenosphere and now overlaid by the continental crust. The dynamics of this body can be controlled either by the rapid southwest motion relative to the Black Sea plate of the plate containing the Carpathians region, or by gravitational sinking into the asthenosphere of an oceanic slab detached from the continental lithosphere (FUCHS et al., 1978). A variant of this model is given by ONCESCU (1984) who assumes that the intermediate-depth earthquakes in Vrancea are not generated inside the subducting lithospheric fragment but in the separation zone, between the fragment and the rest of the subplate,

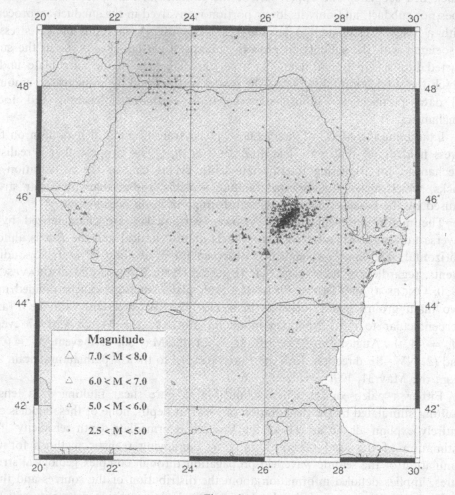

Figure 3
Map of Vrancea seismicity, 1932.1.1–1998.8.1.

which is roughly vertical. CONSTANTINESCU and ENESCU (1984) assume a pale-osubduction from SE to NW and describe the evolution of the region that now corresponds to the Eastern Carpathians, with special reference to the Vrancea region, since the beginning of the consumption of the ancient oceanic plate lying between the Eurasian Plate, the African Plate and Arabian Peninsula).

ENESCU and ENESCU (1993) formulate the hypothesis of an active subduction ongoing in the region, within the area of the Carpathians' continental-type arc. This process, that started 2–3 million years ago, was caused by a slow northwestward movement of the subscrustal lithosphere lying between the Peceneaga-Camena and the Intramoesian faults whereas the crustal lithosphere has been involved in underthrusting motions. The slow movement of the lithosphere strip between the two major faults is likely to be caused by the Anatolian subplate thrusting on the Black Sea subplate. The upper part of the continental lithosphere, being lighter, does not subduct, and only its lower portion is involved in the subduction process, with a velocity estimated to be about 5.0 cm/year. As a result of the stresses associated with the subduction process, subcrustal earthquakes occur in the sub-ducted lithosphere fragment and in a surrounding area in the microplate under which the subduction takes place. The model is supported by an increased amount of data pertinent to lithospheric structure, earthquake location and focal mechanism.

Examining the effects of viscous flow, phase transition and dehydration on the stress field of the relic slab, ISMAIL-ZADEH *et al.* (1999) propose that a realistic mechanism for triggering intermediate-depth events can be the dehydration of rocks, which makes fluid-assisted faulting possible, rather than the shear stress caused by the basalt-ecoglite phase transformation in an oceanic slab.

The major intermediate-depth Vrancea earthquakes are characterized by a reverse faulting mechanism with the T-axis almost vertical and the P-axis almost horizontal. The same mechanism is observed for more than 70% of the studied events, regardless of their magnitude (ENESCU, 1980; ENESCU and ZUGRĂVESCU, 1990; ONCESCU and TRIFU, 1987). The fault plane solutions can be divided into two main groups mainly oriented in a: (1) NE–SW direction, with the P-axis perpendicular to the Carpathian mountain arc (e.g., the March 4, 1977 event, $M_w = 7.4$, the August 30, 1986 event, $M_w = 7.1$, the May 30, 1990 event, $M_w = 6.9$); and (2) NW–SE direction, with the P-axis parallel to the Carpathian mountain arc (e.g., the May 31, 1990 event, $M_w = 6.4$).

Either a paleo- or an active subduction, a pure shear faulting or a tensile faulting stimulated by the dehydration of rocks at depth, none of these models can entirely explain all the aspects of the Vrancea intermediate-depth seismicity. The estimation of the local seismic hazard, by employing realistic methods for the simulation of the seismic wavefield propagation through complex geological struc-tures, implies detailed information about the distribution of the sources and their fault plane solutions, and the database of the Vrancea region seismicity satisfies these requirements.

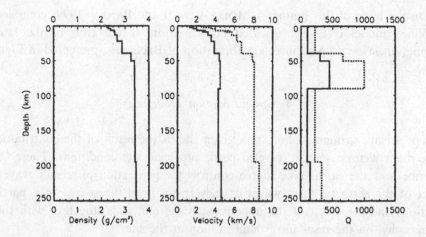

Figure 4

Bedrock structure. Variations with depth of density (in g/cm³), P- and S-wave phase velocity (in km/s), quality factor, Q, for P- (continuous line) and S-wave (dotted lines), in the uppermost 250 km.

3. Structural Models Considered for the Seismic Wavefield Modelling

The ground motion in the Bucharest area is simulated considering an averaged regional (bedrock) model for the Vrancea-Bucharest path, and a local, laterally varying, anelastic model for the sedimentary setting of the city.

The bedrock structure, shown in Figure 4 down to a depth of 250 km, is compiled after RADULIAN *et al.* (1996b) considering: (a) for the crust, the velocity model used for event location with the Romanian telemetered observatories, and (b) for the deeper structure, a low-velocity channel from 90 to 190 km with standard Q values. Below the depth of 250 km, an average continental model is adopted. To investigate the influence of V_S and Q variations within reasonable limits, four variants of the bedrock structure have been considered. V_S changes affect significantly only arrival times of the signals, and Q variations do not produce relevant changes in the simulated waveforms.

The sedimentary formation of the Bucharest area consists of alluvium, loess like, gravel, sand, clay and sandy marl. The presence of unconsolidated sediments (deep soft soils) with irregular geotechnical characteristics and distribution in space was detected by different civil construction enterprises (e.g., "Proiect Bucuresti" Institute, S.C. "Prospectiuni" S.A., "Metrou" S.A.), that have made available a large amount of geological, geotechnical and hydrogeological data (the geotechnical bore-holes alone exceed 10,000). In this framework, more than 2000 bore-holes were analyzed, and the seismic wave velocity was measured by seismic refraction in more than 200 points. We use the synthesis of these results given by MÂNDRESCU and RADULIAN (1999), and we determine the quality factors from empirical correlation with geology, and from similar data published in the literature.

On the basis of this synthesis, MOLDOVEANU and PANZA (1999) compiled the simplified model (NE–SW oriented cross section of the city) of the laterally varying, anelastic deep sedimentary formation of Bucharest, presented in Figure 5.

4. Ground Motion Modelling

The seismic ground motion, at a given site, is the result of the contribution of three main factors: source, traveled path, and local site conditions. These factors describe how the earthquake source controls the generation of seismic waves, the effect of the earth on these waves as they travel from the source to a particular location, and the effect of the uppermost rocks and soils, together with the site topography, on the resultant ground motion at the site.

The simulation of the ground motion is performed using a complex hybrid method (FÄH, 1991; FÄH and PANZA, 1994; FÄH *et al.*, 1994) that combines the modal summation technique (PANZA, 1985; VACCARI *et al.*, 1989; FLORSCH *et al.*, 1991; ROMANELLI *et al.*, 1996), used to describe the *SH* and *P-SV*-wave propaga-

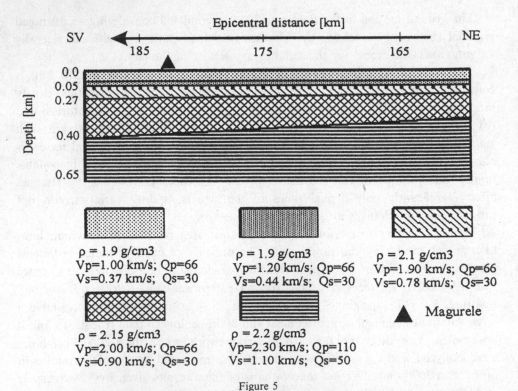

Figure 5
Simplified local structure used for the ground motion modelling in Bucharest. The position of Magurele station is indicated by a full triangle.

Table 1

Source parameters of the Vrancea events considered as scenario earthquakes in the simulations

Event	Lat. (N)	Lon. (E)	Depth (km)	Dip	Rake	Strike	M_0 (N·m)	M_w
May 30, 1990 (A)	45.90°	26.81°	74 ± 17	63°	101°	236°	$3.0 \cdot 10^{19}$	6.9
May 30, 1990(*)	45.86°	26.98°	90 ± 1	65–70°	100°	225°	$4.0 \cdot 10^{19}$	6.9
May 31, 1990 (B)	45.67°	26.00°	87 ± 13	69°	106°	309°	$3.2 \cdot 10^{18}$	6.4

Note: (*) the source parameters determined by TRIFU et al. (1990).

tion in the anelastic bedrock structure, with the finite difference technique (ALTERMAN and KARAL, 1968; BOORE, 1972; KELLY et al., 1976). Therefore, the synthetic signals simulated with the hybrid method are complete in a given frequency-phase velocity window, and take into account the effects of the source, path and local geological conditions. We use the same laterally varying structural model adopted by MOLDOVEANU and PANZA (1999). The source is modeled with a scaled point double-couple, the finiteness of the fault being accounted for by scaling the synthetic signals with the empirical source spectra scaling curves (GUSEV, 1983) modified for the Vrancea intermediate-depth events (MOLDOVEANU et al., 1999). The frequency range covered by the simulations is 0.005–1.0 Hz and allows us the modelling of the seismic input appropriate for 10-story and higher buildings. The investigated frequency window is in agreement with the observed predominant period 1.0–1.5 s of the ground motion induced by the major Vrancea subcrustal earthquakes in Bucharest (MÂNDRESCU and RADULIAN, 1999).

The two scenario earthquakes used for investigating the variation of the local soil effects in Bucharest are representative of the intermediate-depth Vrancea seismicity. These are the earthquakes of May 30, 1990 ($M_w = 6.9$) and May 31, 1990 ($M_w = 6.4$). The details of the source parameters, reported in the CMT catalogue (DZIEWONSKI et al., 1991), are presented in Table 1. The source parameters of the May 30, 1990, Vrancea event determined by TRIFU et al. (1990) (also indicated in Table 1, (*)) using aftershock activity and inversions of the Rayleigh and P waves are in good agreement with the parameters reported in the CMT catalogue. MOLDOVEANU and PANZA (1999) chose the source parameters for modelling the seismic ground motion in Bucharest due to the Vrancea earthquake of May 30, 1990, spanning a wide range of different sets of these values, including those determined by TRIFU et al. (1990) and the CMT catalogue. The hypocentral depth of the source represents one key parameter of the seismic wavefield simulation. Using the available records from Magurele station of the May 30, 1990, Vrancea event and varying the source depth from 10 to 100 km, MOLDOVEANU and PANZA (1999) determine the focal depth by constraining the relative ratios among the components of the simulated signals to reproduce the observed values. The focal depth so determined is $H = 60$ km, compatible, within the error limits, with the CMT depth determination. The shape, peak ground acceleration (PGA) and ratios between the components of the simulated signals are almost unchanged when the rake angle varies from 91° to 111°. Significant variation of the shape and PGA are caused by the variation of the deep angle in the extreme range from 53° to 73°, the most sensible component to this source paramenter variation being the radial one. As a result of these tests, MOLDOVEANU and PANZA (1999) decided to use the CMT catalogue since it supplies a better reproduction of the available records. For internal consistency, in this study we prefer the fault plane parameters of May 31, 1990, Vrancea event reported in the CMT catalogue.

Starting from the results of MOLDOVEANU and PANZA (1999), we consider the May 31, 1990 ($M_w = 6.4$) Vrancea event as earthquake scenario B. The source parameters of this event differ from those of the May 30, 1990, earthquake scenario A, mainly in the azimuth of the fault plane and the focal depth (see Table 1). For both earthquake scenarios, A and B, we consider the hypocentral depth of 60 km, rake angle 101° and the deep angle 63°, and we model the source process with a scaled double-couple. Stability tests concerning the shape, peak ground acceleration (PGA) and ratios between the components of the simulated signals indicate insignificant changes when the rake angle varies from 101° to 106°, and the dip angle varies from 63° to 69°, respectively.

Typical ground motion related quantities, used in seismic engineering for evaluating the local response, are: (a) the peak ground acceleration (PGA), (b) the peak ground velocity (PGV), and (c) the quantity W defined as:

$$W = \lim_{t \to \infty} \int_0^t [(x\tau)\, d\tau]^2\, dt$$

where $x(\tau)$ is the time series describing the ground displacement. Since the ground motion modelling technique we use allows us the computation of a wavefield that contains all the main body- and surface-wave phases, both for *SH*- and *P-SV* motion, the synthetic signals can be processed as the observed time series. It is therefore convenient to consider the ratios PGA(2D)/PGA(1D) and W(2D)/W(1D), i.e., the relative PGA and W, where 2D indicated the computations for the laterally varying model, while 1D represents the computations for the bedrock model. In the spectral domain, suitable quantities to describe the local effects are represented by the relative Fourier transform (spectral ratios) FT(2D)/FT(1D), and the relative spectral amplification Sa(2D)/Sa(1D), considering the response spectra Sa without and with critical damping (e.g.: 5% and 10% of critical damping).

The modelling performed with source A provides results that are in good agreement with the recorded data. Moreover, the peak ground acceleration to peak ground velocity ratios, PGA/PGV, given in Table 2, calculated for each of the three components (radial *R*, vertical *V*, and transversal *T*), both for the synthetic and the observed signals (Magurele station, low-pass filtered with a cutoff frequency of 1 Hz) are in very good agreement with the value determined from globally available strong motion records for deep soft soils—PGA/PGV = 5.0 ± 2.6 (s^{-1})—by DE-CANINI and MOLLAIOLI (personal communication, 1998), and with the value reported earlier by SEED and IDRISS (1982).

5. Seismic Input Simulations

The simulation of the seismic wavefield for frequencies as high as 1 Hz allows us to consider a rather simple geological model of the local structure of Bucharest (Fig. 5). The sampling of the medium for the finite difference scheme is controlled by the lowest value of the S-wave velocity in the model and it is determined equal to 0.06 km. The points where synthetic signals are computed are selected according to the following criteria: (1) sites for which observations and/or records are available, and (2) the difference between sites is an integer multiple of the grid sampling value (0.6 km)

Considering the two different earthquake scenarios A and B, we simulate both acceleration and velocity time series for an array of 35 equally spaced (at·0.6 km) sites located along the previously described profile of Bucharest. Most of the strong motion seismology used accelerograms however, at present, increasing use is made of velocity records, since they supply a measure of motion directly related to kinetic energy. Thus far, relatively little use has been made of displacement time histories, usually limited to the definition of seismic input for seismic isolated structures (e.g., PANZA et al., 1995). The records for a subset of seven sites (the distance between two successive sites is 1.8 km), are illustrated in Figure 6, both accelerations (Figs. 6 (a,b)) and velocities (Figs. 6 (c,d)), corresponding to sources A and B. The signals correspond to a seismic moment $M_0 = 3.0 \cdot 10^{19}$ N · m. The epicentral distances of the first and the last site in Figure 6 are 173.5 and 183.5 km, respectively. The sixth trace from the top is computed for an epicentral distance of 182 km, which is the Magurele station location.

Table 2

PGA/PGV ratios corresponding to the three components—radial (R), vertical (V) and transversal (T)—of the observed and simulated signals for Magurele station (MOLDOVEANU and PANZA, 1999). For deep soft soils the globally available value is PGA/PGV = 5.0 ± 2.6 (s⁻¹).

	R (s^{-1})	V (s^{-1})	T (s^{-1})
Magurele–observed	3.1	3.9	3.4
Magurele–synthetic (event A)	2.9	4.6	4.0

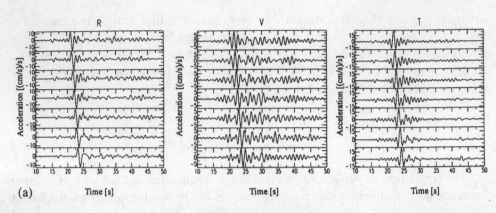

Figure 6a

Acceleration time series (in cm/s²) computed for a set of seven successive receivers spaced by 1.8 km; source *A*; *R*—radial component, *V*—vertical component, *T*—transversal component; seismic moment $M_0 = 3.0 \cdot 10^{19}$ N · m; frequency range 0.005–1.0 Hz. The epicentral distances for the first and the last trace are 173.5 and 183.5 km, respectively. The sixth trace from the top corresponds to Magurele station location.

The comparison of the two sets of signals is performed using: (a) the maximum amplitudes of the signals (peak ground acceleration—PGA, and peak ground velocity—PGV, (b) the shape and (c) the total duration of seismograms. The radial (*R*) and the vertical (*V*) components vary with the changing focal mechanism, while the transversal (*T*) component is rather stable.

The results of the computation of the ratios (PGA_B/PGA_A and PGV_B/PGV_A, where the index *A* or *B* indicates the earthquake scenario, are summarized in Table 3. The *V* component is affected by the change of the focal mechanism, its maximum amplitude for event *B* (PGA_B and PGV_B) increases by a factor of 2 with respect to event *A* (PGA_A and PGV_A), while the simulated waveforms do not change

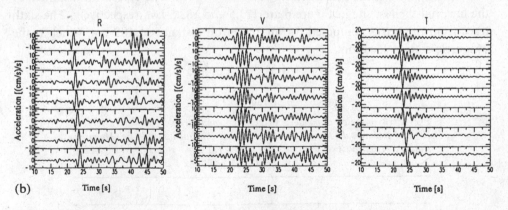

Figure 6b

Acceleration time series (in cm/s²) computed for a set of seven successive receivers spaced by 1.8 km; source *B*.

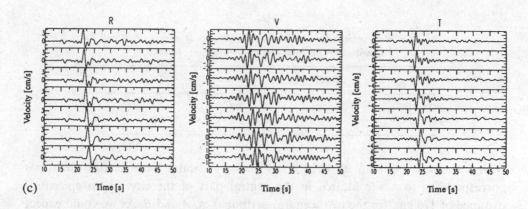

Figure 6c
Velocity time series (in cm/s) computed for a set of seven successive receivers spaced by 1.8 km; source
A. R—radial component, *V*—vertical component, *T*—transversal component.

significantly. The *R* component has smaller amplitude variation, maximum 1.5, and similar waveforms, while the *T* component is quite stable, both in the peak values and waveform shape. The duration of the synthetic *R* and *V* components (both acceleration and velocity) increases with about 5 s in the case of event *B* with respect to event *A*.

The spatial variation along the local profile of Bucharest of the relative quantities PGA(2D)/PGA(1D) and *W*(2D)/*W*(1D) that we obtain considering the two sources, *A* and *B*, are shown in Figure 7. The main changes in the spatial distribution of the relative PGA and *W* are observed in the *R* and *V* components, while the *T* component is very stable with varying mechanism.

Similar stability analysis can be performed in the frequency domain, considering the undamped relative spectral amplifications and the spectral ratios, smoothed

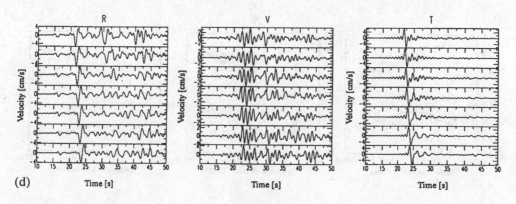

(d)

Figure 6d
Velocity time series (in cm/s) computed for a set of seven successive receivers spaced by 1.8 km; source
B.

Table 3

*The ratios of the peak values of the three components of motion (R, V and T)
simulated for the two scenario earthquakes A and B*

	R	V	T
PGA_B/PGA_A	1.5	2.0	1.3
PGV_B/PGV_A	1.0	2.0	1.0

with a frequency window of 0.025 Hz. Figure 8 indicates these relative quantities corresponding to a site located in the central part of the city, at an epicentral distance of 176 km, for the two scenario earthquakes, *A* and *B*. As we could expect from Figure 7, the behavior of Sa(2D)/Sa(1D) varies significantly for the *R* and *V* components, while for the *T* component it is quite stable. The maximum values of the two relative spectra are in the same range for all three components of motion, both for *A* and *B* earthquake scenarios.

The position of the peaks is different for the different components, and changes for the radial (*R*) and vertical (*V*) components, with the source considered, *A* and *B*. The large excitation of the radial component (2.5 times for source *A* and 3.8 times for source *B*) for the frequency around 0.35 Hz, is not seen in the vertical component that, in the case of source *A*, has mainly four peaks of relative values greater than 2.0 for the frequencies around 0.53, 0.65, 0.75 and 0.95 Hz, and three peaks of relative values greater than 2.0 for the frequencies around 0.5, 0.75 and 0.9 Hz, in the case of source *B*. The transverse component has two peaks of relative values, around 0.42 and 0.9 Hz, greater than 2.5 for source *A*, and greater than 2.25 for source *B*, respectively. The resonance frequency of the sedimentary layers might explain these spectral amplifications at the considered site. For example, the peak around 0.4 Hz can be due to the resonance of the upper 400 m, while the peak of

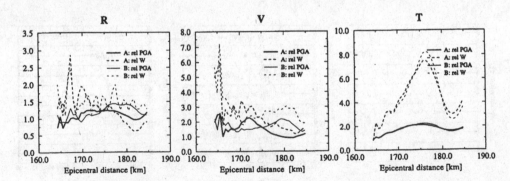

Figure 7

Sources A and B: Spatial distribution of the relative PGA (PGA(2D)/PGA(1D))—continuous line, and relative *W* (*W*(2D)/*W*(1D))—dotted line, for the *R*, *V*, and *T* components of motion along the cross section. 2D stands for the local sedimentary structure, while 1D stands for the bedrock structure.

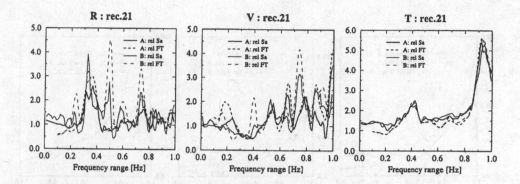

Figure 8

Sources A and B: Relative response spectra Sa(2D)/Sa(1D) for zero damping—continuous line, and spectral ratio FT (2D)/FT(1D) for 0.025 Hz smoothing—dotted line, obtained at a receiver located in the center of Bucharest, at the epicentral distance of 176 km; R, V, and T identify the components of motion.

0.9 Hz can be due to the resonance of the uppermost 270 m. The maximum values of the quantities describing the ground motion are given in Table 4, for each of the three components and for both scenario earthquakes.

Figures 7 and 8 and Table 4 we can observe that: (1) the local response along the profile is sensible to the seismic source, the most affected begin the vertical (V) and the radial (R) components, (2) the maximum values of the relative PGA and PGV change by no more than one unit, as well as the spectral quantities, (3) the computed local effect, expressed as the ratios PGA/PGV, is in very good agreement with the values reported for the deep soft soils (DECANINI and MOLLAIOLI, personal communication, 1998; SEED and IDRISS, 1982).

For engineering purposes, substantially used it is the response spectrum, Sa, computed for different critical damping values. In Figures 9 (a,b) we display these response spectra for the three components of motion, computed with a fraction of

Table 4

Peak values obtained along the profile in Bucharest (W, PGA and PGA/PGV), and spectral values (Sa and FT) corresponding to a site located in the center of the city, at an epicentral distance of 176 km.

Component	R		V		T	
Source	A	B	A	B	A	B
W(2D)/W(1D)	2.7	2.5	6.0	7.0	7.9	8.2
PGA(2D)/PGA(1D)	1.3	1.5	2.3	2.3	2.0	2.0
PGA/PGV	3.3	3.9	4.8	4.8	4.7	4.7
Sa(2D)/Sa(1D)	2.6	3.9	3.5	3.2	5.5	5.3
FT(2D)/FT(1D)	2.7	4.6	4.2	4.2	5.0	5.5

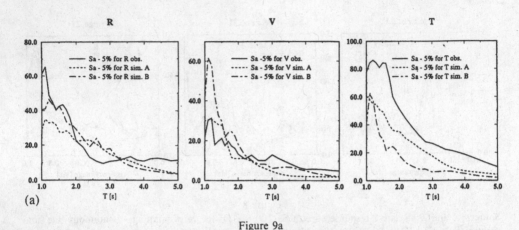

(a)

Figure 9a

Response spectra (Sa in cm/s²) for 5% of critical damping, corresponding to the epicentral distance of 182 km both for the May 30, 1990 Vrancea event, accelerograms recorded at Magurele station—continuous line, and for the synthetic signals corresponding to source *A*—dotted line, and to source *B*—long dashed line. *R*, *V*, and *T* identify the three ground motion components.

critical damping equal to 5% and 10% respectively, both for the accelerograms recorded at Magurele station for the May 30, 1990 Vrancea event, and for the two different sets of simulated signals (earthquake scenarios *A* and *B*). Even if a simplified source process is considered in the simulations, the synthetic signals reproduce most of the main features of the observations that are relevant for seismic engineering.

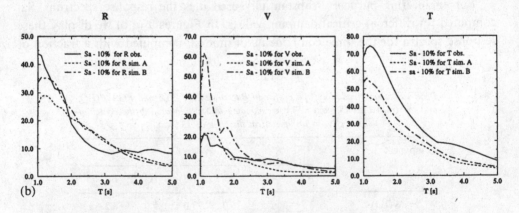

(b)

Figure 9b

Response spectra (Sa in cm/s²) for 10% of critical damping, corresponding to the epicentral distance of 182 km both for the May 30, 1990 Vrancea event, accelerograms recorded at Magurele station—continuous line, and for the synthetic signals corresponding to source *A*—dotted line, and to source *B*—long dashed line. *R*, *V*, and *T* identify the three ground motion components.

5. Conclusions

The mapping of seismic ground motion due to the events originating in a given seismogenic zone can be made by measuring the seismic signals with a dense set of recording instruments when a strong earthquake occurs or/and by computing theoretical signals, using the available information regarding tectonic and geological/geotechnical properties of the medium where seismic waves propagate. Strong earthquakes are very rare phenomena and this makes it very difficult (practically impossible in the near future) to prepare a sufficiently large database of recorded strong motion signals that could be analyzed in order to define generally valid ground parameters, to be used in seismic hazard estimations.

While waiting for the increment of the strong motion data set, a very useful approach to perform immediate microzonation is the development and use of modelling tools based, on one hand, on the theoretical knowledge of the physics of the seismic source and of wave propagation and, on the other hand, exploration of the rich database of geotechnical, geological, tectonic, seismotectonic, and historical information already available.

Using a scaled point double-couple source model and relatively simple path (bedrock) and local structure models, MOLDOVEANU and PANZA (1999) succeeded in reproducing, for periods greater than 1 s, the available recorded ground motion in Bucharest (from Magurele station), at a very satisfactory level for seismic engineering.

Parametric tests that represent a major advantage of the numerical simulations, a powerful and economically valid tool for seismic microzonation, have been performed considering the two fault plane solutions representative of a major Vrancea intermediate-depth earthquakes. The two scenario earthquakes considered in this study are representative of the major events which occurred in this region. The site effects simulated in the frequency range up to 1 Hz indicate a quite stable behavior for the T component, while the R and V components are sensitive to the scenario earthquake. Therefore, although the strongest local effect is measured (both observed and synthetic) in the T component, for a reliable determination of the seismic input all three components of motion (R, V, T) should be used.

Acknowledgements

The authors have been supported by: (1) Copernicus project CIP-CT94-0238 "Quantitative Seismic Zoning of the Circum-Pannonian region," (2) NATO linkage grant AS.12-2-02 (ENVIR.LG 960916) 677(96) LVdC "Microzonation of Bucharest, Russe and Varna in connection with Vrancea earthquakes," (3) UNESCO-IGCP project 414 "Seismic Ground Motion in Large Urban Areas." One of the Authors (CLM) is grateful to the "Consorzio per lo Sviluppo Internazionale,"

Universita' di Triste, for awarding a one-year scholarship at Departimento di Scienze della Terra. Many thanks are extending to Dr. K. Atakan and Dr. C-I. Trifu for their careful review of the manuscript.

Realistic Modelling of Seismic Input for Megacities and Large Urban Areas (project 414)

REFERENCES

ALTERMAN, Z. S., and KARAL, F. C. (1968), *Propagation of Elastic Waves in Layered Media by Finite Difference Methods*, Bull. Seismol. Soc. Am. *58*, 367–398.

BOORE, D. M., *Finite difference methods for seismic waves propagation in heterogeneous materials*. In *Methods in Computational Physics* (ed. Bolt, B. A.), vol. 11 (Academic Press, New York 1972) pp. 1–37.

CONSTANTINESCU, L., and ENESCU, D. (1984), *A Tentative Approach to Possibly Explaining the Occurrence of the Vrancea Earthquakes*, Rev. Roum. Geol. Geogr. Geophys. *28*, 19–32.

DECANINI, L., and MOLLAIOLI, F. (1998), personal communication.

DZIEWONSKI, A. M., EKSTRÖM, G., WOODHOUSE, J. H., and ZWART, G. (1991), *Centroid-moment Tensor Solutions for April–June 1990*, Phys. Earth Planet. Inter. *66*, 133–143.

ENESCU, D. (1980), *Contributions to the Knowledge of the Focal Mechanism of the Vrancea Strong Earthquake of March 4, 1977*, Rev. Roum. Géol. Géophys. Géogr. Géophys. *24*, 3–18.

ENESCU, D., and ZUGRĂVESCU, D. (1990), *Geodynamic Considerations Regarding the Eastern Carpathians Arc Bend, Base on Studies on Vrancea Earthquakes*, Rev. Roum. Géophysique *34*, 17–34.

ENESCU, D., and ENESCU, B. D. (1993), *A New Model Regarding the Subduction Process in the Vrancea Zone*, Rom. J. Phys. *38*, 321–328.

FÄH, D. (1991), *Stima del moto sismico del suolo in bacini sedimentari*, Tesi di dottorato, tutor: G. F. Panza, Trieste University.

FÄH, D., and PANZA, G. F. (1994), *Realistic Modelling of Observed Seismic Motion in Complex Sedimentary Basins*, Annali di Geofisica, *XXXVII*, 6, 1771–1796.

FÄH, D., SUHADOLC, P., MUELLER, ST., and PANZA, G. F. (1994), *A Hybrid Method for the Estimation of the Ground Motion in Sedimentary Basins: Quantitative Modelling for Mexico City*, Bull. Seismol. Soc. Am. *84* (2) 383–399.

FLORSCH, N., FÄH, D., SUHADOLC, P., and PANZA, G. F. (1991), *Complete Synthetic Seismograms for High-frequency Multimode SH-waves*, Pure appl. geophys. *136*, 529–560.

FUCHS, K., BONJER, K. P., BOCK, G., RADU, C., ENESCU, D., JIANU, D., NOURESCU, A., MERKLER, G., MOLDOVEANU, T., and TUDORACHE, G. (1978), *The Romanian Earthquake of March 4, 1997. Aftershocks and Migration of Seismic Activity*, Tectonophysics *53*, 225–247.

GUSEV, A. A. (1983), *Descriptive Statistical Model of Earthquake Source Radiation and its Application to an Estimation of Short-period Strong Motion*, Geophys. J. R. Astr. Soc. *74*, 784–808.

ISMAIL-ZADEH, A. T., PANZA, G. F., and NAIMARK, B. M. (2000), *Stress in the Descending Relic Slab beneath Vrancea, Romania*, Pure appl. geophys. *157*, 111–130.

KELLY, K. R., WARD, R. W., TREITEL, S., and ALFORD, R. M. (1976), *Synthetic Seismograms: A Finite Difference Approach*, Geophysics *41*, 2–27.

MCKENZIE, D. P. (1970), *Plate Tectonics of Mediterranean Region*, Nature *226*, 239–242.

MCKENZIE, D. P. (1972), *Active Tectonics of the Mediterranean Region*, Geophys. J. R. Astr. Soc. *39*, 109–185.

MÂNDRESCU, N., and RADULIAN, M., *Seismic microzoning of Bucharest (Romania): A critical review*. In *Vrancea Earthquakes: Tectonics, Hazard, and Risk Mitigation* (eds. Wenzel, F., and Lungu, D.) (Kluwer Academic Publishers 1999) pp. 109–122.

MOLDOVEANU, C. L., KUZNETSOV, I., NOVIKOVA, O. V., PANZA, G. F., and VOROBIEVA, I. A. (1998), *Monitoring of the Preparation of the Future Strong Earthquakes in Vrancea, Romania, Using the CN Algorithm*, ICTP preprint, IC/98/29, Trieste, Italy.

MOLDOVEANU, C. L., and PANZA, G. F., *Modelling, for microzonation purposes, of the seismic ground motion in Bucharest due to the Vrancea earthquake of May 30*. In *Vrancea Earthquakes: Tectonics, Hazard, and Risk Mitigation* (eds. Wenzel, F., and Lungu, D.) (Kluwer Academic Publishers 1999) pp. 85–98.

MOLDOVEANU, C. L., RADULIAN, M., PANZA, G. F., and VACCARI, F. (1999), *Scaling of the Subcrustal Events in the Vrancea Region (Romania)*, in preparation.

ONCESCU, M. C. (1984), *Deep Structure of the Vrancea Region, Romania, Inferred from Simultaneous Inversion for Hypocenters and 3-D Velocity Structure*, Ann. Geophys. *2*, 23–28.

ONCESCU, M. C., and TRIFU, C.-I. (1987), *Depth Variation of the Moment Tensor Principal Axes in Vrancea (Romania) Seismic Region*, Ann. Geoyphysicae *5B*, 149–154.

ONCESCU, M. C., MÂRZA, V. I., RIZESCU, M., and POPA, M., *The Romanian earthquake catalogue between 984–1996*. In *Vrancea Earthquakes: Tectonics, Hazard, and Risk Mitigation* (eds. Wenzel, F., and Lungu, D.) (Kluwer Academic Publishers 1999) pp. 43–48.

RADULIAN, M., MÂNDRESCU, N., and PANZA, G. F. (1996a), *Seismogenic Zones for Romania*, ICTP preprint, IC/96/255, Trieste, Italy.

RADULIAN, M., ARDELEANU, L., CAMPUS, P., ŠILENY, J., and PANZA, G. F. (1996b), *Waveform Inversion of Weak Vrancea (Romania) Earthquakes*, Studia Geoph. et Geod. *40*, 367–380.

RADULIAN, M., MÂNDRESCU, N., PANZA, G. F., POPESCU, E., and UTALE, A. (2000), *Characterization of Romania Seismogenic Zones*, Pure appl. geophys. *157*, 57–77.

ROMANELLI, F., BING, Z., VACCARI, F., and PANZA, G. F. (1996), *Analytical Computations of Reflection and Transmission Coupling Coefficients for Love Waves*, Geophys. J. Int. *125*, 123–138.

PANZA, G. F. (1985), *Synthetic Seismograms: The Rayleigh Waves Modal Summation*, J. Geophys. *58*, 125–145.

PANZA, G. F. (1993), *Synthetic Seismograms for Multimode Summation—Theory and Computational Aspects*, Acta Geod. Geoph. Mont. Hyng. *28* (1–2), 197–247.

PANZA, G. F., VACCARI, F., COSTA, G., SUHADOLC, P., and FÄH, D., *Seismic input modelling: a key issue for the safe implementation of seismic isolation*. In *Proceedings of the International Post-SMIRT Conference Seminar on Seismic Isolation, Passive Energy Dissipation and Control of the Vibration of Structures*, Santiago 1995, 47–78.

SEED, H. B., and IDRISS, I. M. (1982), *Ground motions and soil liquefaction during earthquakes; Monograph Series: Engineering Monographs on Earthquake Criteria, Structural Design, and Strong Motion Records*, Coordinating editor, M. S. Agbabian, 134 pages.

TRIFU, C-I., DESCHAMPS, A., RADULIAN, M., and LYON-CAEN, H. (1990), *The Vrancea earthquake of May 30, 1990: an estimate of the source parameters*, XXII General Assembly ESC, Barcelona-1990, Proceedings and Activity Report 1988–1990, 449–454.

VACCARI, F., GREGERSEN, S., FURLAN, M., and PANZA, G. F. (1989), *Synthetics Seismograms in Laterally Heterogeneous, Anelastic Media by Modal Summation of the P-SV Waves*, Geophys. J. Int. *99*, 285–295.

(Received June 9, 1998, revised February 2, 1999, accepted May 12, 1999)

Pure appl. geophys. 157 (2000) 269–279
0033–4553/00/020269–11 $ 1.50 + 0.20/0

Pure and Applied Geophysics

The Dependence of Q with Seismic-induced Strains and Frequencies for Surface Layers from Resonant Columns

GH. MARMUREANU,[1] D. BRATOSIN[2] and C. O. CIOFLAN[1]

Abstract—The gross effect of internal friction is summarized by the dimensionless quantity Q, defined in various ways. If a volume of soil is cycled in stress at a frequency ω, physically, the Q factor is equal to the ratio of energy dissipated per cycle to the total energy $Q^{-1} = \Delta E/(2\pi E)$. The authors used Hardin and Drnevich resonant columns to determine the damping capacity of cylindrical specimens from surface soil layers during torsional and longitudinal vibrations. The energy dissipated by the system is a measure of the damping capacity of the soil. The damping will be defined by the shear damping ratio for the soil D, analogous to the critical viscous damping ratio for a single degree of freedom c/c_0. Values of damping determined in these resonant columns will correspond to the area of the hysteresis loop stress strain relation divided by 4π times the elastic strain energy stored in the specimen at maximum strain. Consequently, we can express D in the form of quality factor Q, that is $Q = 1/(2D)$, where Q is defined in terms of the fractional loss of energy per cycle of oscillation and D is a nonlinear function ω and γ. The nonlinear dependence of Q with seismic induced strains and frequencies for large deformations has an important influence on the propagation of the seismic waves in the hazard and microzonation studies.

Key words: Quality factor, damping ratio, internal friction, resonant columns, seismic hazard.

Introduction

In real materials, wave amplitude is attenuated as a result of a variety of processes, inclusively the scattering; we can summarize macroscopically as internal friction. The gross effect of the internal friction is taken into account by the dimensionless quantity Q, defined in various ways. If a volume of soil is cycled in stress at a frequency ω, physically, the Q factor is equal to the ratio of energy dissipated per cycle to the energy:

$$\frac{1}{Q} = \frac{1}{2\pi} \cdot \frac{\text{energy dissipated in one cycle } (\Delta E)}{\text{peak energy stored during the cycle} (E)} \tag{1}$$

[1] National Institute for Earth Physics, Magurele-Bucharest, 76900, Romania. Tel/fax: +40-1-4930118.
[2] Institute of Solid Mechanics, Romanian Academy, 15 C. Mile, Bucharest, Romania.

From classical observations, there are results which show a dependence of Q with angular frequency ω (SIPKIN and JORDAN, 1979). PING LI *et al.* (1995), in a special case of in-seam seismic exploration, found that the attenuation coefficient Q_α for Love-type channel wave is a nonlinear function of angular frequency.

Nonlinearity of soil response is also called strain dependence, because the strain level goes through during an earthquake increases with the level of stress or ground motion. The question of the level at which nonlinearity becomes significant is of great relevance in hazard estimation in that very often the more frequency low-level ground motions, such as those recorded during small earthquakes, are used to estimate site response during large earthquakes. The purpose of this study is to ascertain if the attenuation coefficient Q is a function of angular frequencies and induced strains. The authors used the methodology and the methods of test for shear modulus and damping ratio of soils by the Hardin and Drnevich resonant columns.

Phenomenological Model for Seismic Attenuation

The simplest descriptions of attenuation can be developed by considering a mechanical linear oscillator (Fig. 1) of mass m and a spring with spring constant k. The model has a single degree of freedom and the attenuation is introduced by adding a damping force f, proportional to the velocity, $c \cdot x(t)$, as a friction between the moving mass and the underlying surface.

In the case of "source-free motion" (also called transient, natural, homogeneous, complementary) the equation of motion can be written as:

$$x(t) + 2\alpha\omega_0 \cdot x(t) + \omega_0^2 \cdot x(t) = 0 \qquad (2)$$

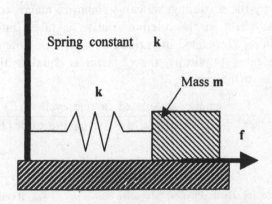

Figure 1
The phenomenological model for seismic attenuation.

where $k/m = \omega_0^2$; $c_0/m = 2\alpha\omega_0$; where c_0 represents the critical viscous damping coefficient and α is the coefficient of friction (dimensionless; if $\alpha = 0$, no attenuation). The value of the damping coefficient for an $\omega \neq \omega_0$, known as angular frequency of perturbatory force, is:

$$c = 2m\alpha\omega. \tag{3}$$

The ratio between damping coefficient (c) and the critical one (c_0) is a dimensionless parameter named *damping ratio or fraction of critical damping* (D):

$$D = \frac{c}{c_0}. \tag{4}$$

The characteristic equation for (2.1): $r^2 + 2\alpha\omega_0 r + \omega_0^2 = 0$ has the roots:

$$r_{1,2} = -\alpha\omega_0 \pm \omega_0\sqrt{\alpha^2 - 1} \tag{5}$$

when $c\pi c_0$, $\alpha\pi 1$ and if we note $\omega_0^2(\alpha^2 - 1) = -\beta^2$ then $r = -\alpha\omega_0 \pm i\beta$, the solution of equation (2) is:

$$x(t) = A_0 \cdot e^{-\alpha\omega_0 t} \cos(\beta t - \varphi) \tag{6}$$

where the argument φ is the phase at $t = 0$ (Fig. 2).

We can express the damping ratio α in a form of the quality factor Q of the system which is a dimensionless parameter (AKI and RICHARDS, 1980; LAY and WALLACE, 1995):

$$\alpha = \frac{1}{2Q} \tag{7}$$

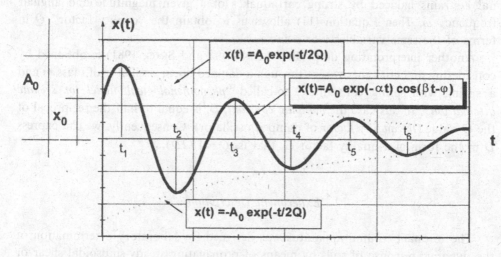

Figure 2
The exponentially decaying amplitude of the oscillator.

and then the amplitude $A(t)$ is a function of time:

$$A(t) = A_0 \cdot e^{-t/2Q} \tag{8}$$

where Q is defined in terms of the *fraction loss* of energy per cycle of oscillation, as is shown in Equation (1). However, the values of $x(t)$ at the corresponding arguments t_n and t_{n+2} are assumed to be close and we may therefore measure the amplitude decay via the *damping ratio*:

$$\left| \frac{x(t_n)}{x(t_{n+2})} \right| \approx \exp\left(-\omega_0 \cdot \frac{T}{2Q} \right) \tag{9}$$

where $T = t_{n+2} - t_n$ is the damping period and $x(t_n)$, $x(t_{n+2})$ are the amplitudes of successive cycles of oscillation (Fig. 2).

One defines *the logarithmic decrement* as:

$$\delta = \ln\left| \frac{x(t_n)}{x(t_{n+2})} \right| = 2\pi D = \frac{\pi}{Q} \tag{10}$$

and from Equations (1) and (10) we can obtain the following connection between the logarithmic decrement δ, the damping ratio D and the quality factor Q:

$$Q = \frac{\pi}{\delta} = \frac{1}{2D}. \tag{11}$$

Using the Hardin and Drnevich resonant columns it is possible to measure the damping ratio D or the logarithmic decrement δ for different shear γ or longitudinal ε strains induced by strong earthquakes of a given magnitude and angular frequency ω. Then Equation (11) allows us to obtain the "quality factor" Q in terms of the same variables γ or ε and ω.

Another interpretation of Q (BEN-MENAHEM and SING, 1981) is obtained by considering an initial energy sent through a damped linear oscillator of mass m and a spring with spring constant k, so-called *"phenomenological model for seismic attenuation."* In this case the *damping constant*, D, is equal to half the reciprocal of the Q-value (*the quality factor* of damped oscillator). Consequently, we can express D in the form of a quality factor Q, that is, $Q = 1/(2D)$.

Experiment Description

The resonant column apparatus was designed for laboratory determination of the dynamic response of soils by means of propagating steady-sinusoidal shear or compressional waves in a cylindrical soil specimen, soil *column*, under *resonant* frequency conditions (Fig. 3).

The sample (identical in shape, dimensions an sampling to those used in the classical triaxial testing: $\Phi = 3.57$ cm, $h = 8$ cm) together with the vibration device, the resonant column system, are enclosed in a cell chamber where the confining conditions (*in situ* conditions) are supplied by compressed air, water or mineral oil. The specimen base is coupled rigidly to the base of cell chamber and the specimen top is attached to the vibration excitation device (fixed-free end conditions).

The torsional vibration exciter consists of four drive coils placed around the specimen top and the longitudinal vibrator is set co-axially with the specimen (see Fig. 3). The oscillating torque and longitudinal force are produced electromagnetically with coils and permanent magnets (current-to-force transducers).

The coil-drive input signal to drive coils is generated by a variable sine wave oscillator and is pre-amplified before input to the drive coils. The signal is also connected to a digital multimeter and the horizontal axis of an oscilloscope display.

The sinusoidal driving torque ($M = M_0 \sin \omega t$) and the longitudinal force ($F = F_0 \sin \omega t$) are directly proportional to the input current, with the torque/current and force/current ratios being apparatus constants.

The motions of the top cap system are measured with accelerometers and the electrical output of each accelerometer is electronically amplified by charge amplifiers, and then connected to the vertical axis of the oscilloscope display and to the same multimeter as the output voltage signal (mV_{rms}). The ratio $mV_{rms}/(m/s^2)$ is a calibration constant.

The excitation frequency is increased from the sine wave generator until resonance occurs, and at this moment one can record the resonant frequency ω_r [Hz], the acceleration A [mV_{rms}] and the current C [mV_{rms}]. The frequency is then set

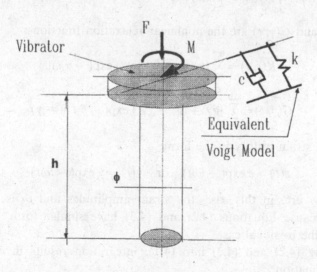

Figure 3
The mechanical model of resonant column.

at $\sqrt{2}$ times the resonant frequency and the acceleration and current are again recorded. These values are used for the determination of the moduli and damping capacities; this method advantageously eliminates the apparatus constants.

Generally, the amplitude of excitation is sufficiently small in order to consider the soil specimen as a linearly elastic specimen uniform in mass, density and dimensions (DRNEVICH *et al.*, 1978).

As a result of this assumption, the irreversible strains vanish and the steady-state of a perturbing excitation induces stationary stress-strain hysteresis loops. Under this condition, the Voigt model can satisfactorily describe the material behavior for a certain constant excitation level. By using this model, with experimental data, one can obtain the elastic moduli (shear G or E modulus), damping (D_t—torsional or D_1—longitudinal) and strain amplitudes (γ—torsional or ε—longitudinal).

Damping Ratio From Resonant Column Test

It was experimentally observed that changing the excitation amplitude modifies the modulus and damping values in terms of new strain levels (DRNEVICH *et al.*, 1978). This nonlinear behavior cannot be explained by the Voigt model and requires the use of an nonlinear viscoelastic model (BRATOSIN, 1993; CHRISTENSEN, 1971). For nonlinear behavior modeling, we assume that soils are nonlinear viscoelastic materials with the constitutive equations:

$$(t) = \int_0^t K(\varepsilon, t-s)\varepsilon(s)\, ds; \quad \tau(t) = \int_0^t G(\gamma, t-s)\gamma(s)\, ds \tag{12}$$

where $K(\varepsilon, t)$ and $G(\gamma, t)$ are the nonlinear relaxation functions:

$$K(\varepsilon, t) = \sum_i^{0,2} [k_i^\infty + (k_i^0 - k_i^\infty)\exp(-\alpha_i t)]\varepsilon^i$$

$$G(\gamma, t) = \sum_i^{0,2} [g_i^\infty + (g_i^0 - g_i^\infty)\exp(-\beta_i t)](-\gamma)^i \tag{13}$$

subjected to a strain-history in the form:

$$\varepsilon(t) = \varepsilon\exp(-i\omega t) \quad \text{or} \quad \gamma(t) = \gamma\exp(-i\omega t) \tag{14}$$

where ε and γ are, in this case, the strain amplitudes and ω is the excitation frequency. Because Equations (4.2) and (4.3) have similar form, we shall next address only the torsional case.

Substituting (4.2) and (4.3) into (4.1), integration results in dynamic shear constitutive equation:

$$(t) = G^*(\gamma, i\omega)\gamma(t) \tag{15}$$

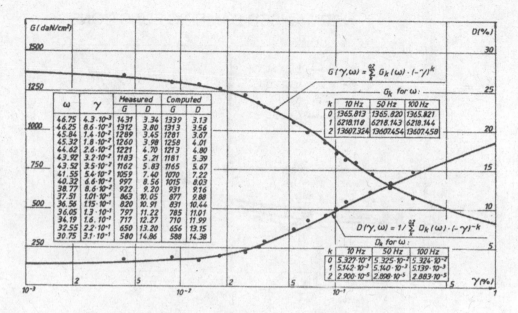

Figure 4
The normalized damping functions for marl at resonant frequencies.

where $G^*(\gamma, i\omega)$ is *the complex modulus function*:

$$G^*(\gamma, i\omega) = G(\gamma, \omega) \exp[i \tan^{-1} D(\gamma, \omega)] \qquad (16)$$

$G(\gamma, \omega)$ being *the dynamic torsional modulus function* and $D(\gamma, \omega)$ is *the torsional damping function*. This last function characterizes the damping properties of the nonlinear material and plays the same role as the damping ratio D from linear dynamics. For example, for marl (Fig. 4), at different resonant frequencies (30.75 Hz–46.75 Hz) we obtained the following damping function.

$$D(\gamma, \omega) = \frac{1}{\sum\limits_{k}^{0,2} D_k(\omega) \cdot (-\gamma)^k} \qquad (17)$$

where D_k for $k = 0$; 1 and 2 are given as in Table 1.

Table 1
The values of D_k for frequencies of 10; 50 and 100 Hz

k	10 Hz	50 Hz	100 Hz
0	$5.327 \cdot 10^{-2}$	$5.325 \cdot 10^{-2}$	$5.324 \cdot 10^{-2}$
1	$5.142 \cdot 10^{-3}$	$5.140 \cdot 10^{-3}$	$5.139 \cdot 10^{-3}$
2	$2.900 \cdot 10^{-5}$	$2.898 \cdot 10^{-5}$	$2.883 \cdot 10^{-5}$

NORMALISED DAMPING FUNCTIONS

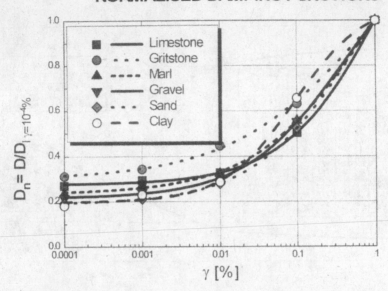

Figure 5
The normalized damping functions at resonant frequencies.

Figure 5 illustrates the presence and levels of nonlinearity of the damping function (normalized), based on several resonant column tests carried out on soil and rock samples. In Figure 6, as an example, a damping function $D(\gamma, \omega)$ obtained in the Drnevich resonant column test performed on a clay sample is given. Finally, using Equation (2.10) one can obtain the factor Q in terms of strain level γ and frequency ω as $Q(\gamma, \omega) = 1/(2\,D(\gamma, \omega))$ and the result is given in Figure 7, where the function $D(\gamma, \omega)$ from Figure 6 was used.

Conclusions

The classic definition of Q, as the ratio of energy dissipated per cycle to the total energy, is rarely of direct use. Since only in special experiments is it possible to drive a material element with stress waves of unchanging amplitude and period. A common hypothesis in seismology which involves attenuation of a signal composed of a range of frequencies, is that the attenuation is a linear phenomenon.

Using the Hardin and Drnevich resonant columns, it is possible to measure the damping ratio D for different shear (γ) or longitudinal (ε) strains induced by strong earthquakes of a given magnitude (strain induced) and frequency ω. The results presented in this paper, using resonant columns (Hardin and Drnevich), indicate a nonlinear dependence of the dynamic modulus and damping function in terms of the strain level (i.e., earthquake magnitude) and frequency. For example, we have *the dynamic torsional modulus* $G = G(\gamma, \omega)$, and *the torsional damping function*, $D = D(\gamma, \omega)$, functions of shear strain, γ, and frequency, ω. This last function characterizes the damping properties of the nonlinear material and plays the same role as the damping ratio, D, for linear dynamics.

It was experimentally observed that changing the excitation amplitude modifies the modulus and damping values in terms of a new strain level. This nonlinear cannot be explained by the Voigt model and necessitates the utilization of a nonlinear viscoelastic model.

Figures 4–7 illustrate the presence and levels of nonlinearity in damping function, D, and attenuation factor, $Q = Q(\gamma, \omega)$, for several resonant columns tests carried out on soil and rock samples representing limestones, gritstones, marls, gravels, sands, clays, and other sedimentary materials.

The experimental results support the conclusion that seismic attenuation in sediments from surface layers is a function of strain levels (earthquake magnitude) induced by strong earthquakes, frequency and other factors which can affect the dynamic response of the sedimentary soils, namely, preexisting spherical stress (σ), water content (w), consolidation time (t_C), density of sample (ρ) etc. (i.e., $Q = Q(\gamma, \omega, \sigma, w, t_C, \rho \ldots).$).

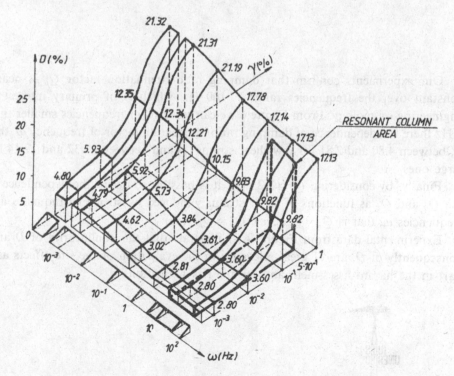

Figure 6

Damping function $D(\gamma, \omega)$ from Drnevich resonant column on clay sample at resonant frequencies.

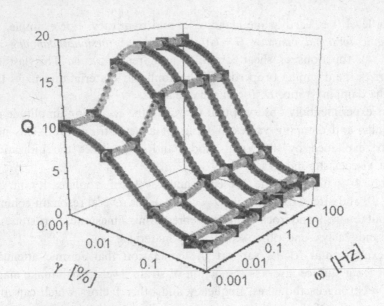

Figure 7

The function $Q_\beta = Q_\beta(\gamma, \omega)$ in terms of strain level γ and frequency ω on a clay sample at resonant frequencies.

Our experiments confirm that damping ratio/attenuation factor Q_β is nearly constant over the frequencies range 7–100 Hz, which is of primary interest in engineering seismology. From Figure 6 we can see that for frequencies smaller than 7 Hz there is a dependence of damping ratio/attenuation factor of frequency ω, that is, between 4.80 and 2.81 for low shear strains (γ) and between 21.32 and 17.14 for large ones.

Finally, by considering $Q_\alpha \cong 2.25\, Q_\beta$, it is possible to derive the dependence of the Q_α and Q_β as functions of shear strains γ induced by strong earthquakes and frequencies ω, that is $Q_\alpha = Q_\alpha(\varepsilon, \omega)$, $Q_\beta = Q_\beta(\gamma, \omega)$.

Experimental data from Figures 4–7 illustrate that the nonlinearity of D and, consequently of Q, are of great relevance in the evaluation of the site effects as a part in the hazard assessment and seismic risk mitigation.

Realistic Modelling of Seismic Input for Megacities and Large Urban Areas (project 414)

REFERENCES

AKI, K., and RICHARDS, P. G., *Quantitative Seismology* (W. H. Freeman and Company, San Francisco 1980).

BEN-MENAHEM, A., and SING, S. J., *Seismic Waves and Sources* (Springer-Verlag, New-York, Inc. 1981).

BRATOSIN, D. (1993), *A Dynamic Constitutive Equation for Soils*, Revue Roumaine des Sciences Technique—Mécanique Appliquée 5.

CHRISTENSEN, R. M., *Theory of Viscoelasticity* (Academic Press, New York 1971).

DRNEVICH, V. P., HARDIN, B. O., and SHIPPY, D. J. (1978), *Modulus and damping of soils by resonant-column method*. In *Dynamic Geotechnical Testing*, ASTM.

LAY, TH., and WALLACE, T. G., *Modern Global Seismology* (Academic Press, Inc. 1995).

PING LI, SCHOTT, W., and RÜTER, H. (1995), *Frequency-dependent Q-estimation of Love-type Channel Waves*, Geophysics 60 (6), 1773–1789.

SIPKIN, S. A., and JORDAN, T. H. (1979), *Frequency dependence of QScS*, Bull. Seismol. Soc. Am. 69, 1055–1079.

(Received April 24, 1998, revised September 3, 1998, accepted April 28, 1999)

To access this journal online:
http://www.birkhauser.ch

Notes to Authors

PAGEOPH welcomes original contributions in English (and occasionally in French and German) on all aspects of geophysics. All manuscripts should be submitted to the Regular Issues Editor-in-Chief, in triplicate, formatted with double spacing and wide margins. For further details see the following paragraphs.

Format of Manuscripts

Length and Page Charges: A paper should not exceed 16 printed pages including tables and figures. For articles exceeding 16 printed pages the authors will be charged sFr. 80.00 for each additional page. No page charges, except those for color prints, are required for contributors to special issues.

Title Page: This should include the the complete title, full names and addresses of all authors. In addition, corresponding authors should provide their fax number and e-mail address if they are available.

Abbreviated Title: It is necessary to indicate an abbreviated title, which will be used as a running head (no more than 50 characters including spaces).

Abstract: The abstract should be in English, and in the language of the text, if different. It has to be of no more than 10 sentences and should be concise and self-contained.

Keywords: Up to 6 keywords should be listed, suitable for incorporation into information-retrieval systems.

Text: The text must include a citation for each item listed under References; the approximate position of each figure should be indicated in the text. The metric system should be used throughout the text, figures, and tables.

Tables: Tables are to be presented on separate pages, with a brief title for each.

Figures: Figure captions and legends are to be typed on a separate page or pages as the last element in the manuscript. Make sure that line thickness and lettering allow an adequate size reduction. Heliographic or photocopies are not suitable for reproduction. Highquality, glossy, photographic prints must be submitted. Color prints are permitted but authors will be charged for them.

References: They are to be listed in alphabetical order in the following style:
Journal article: Haurwitz, B., and Cowley, A.D. (1973), The Diurnal and Semidurnal Barometric Oscillations, Global Distribution and Annual Variation, Pure Appl. Geophys. 102, 193-222.

Whole book: Bath, M., Introduction to Seismology (Birkhäuser, Basel 1973).

Article in a book: Haurwitz, B., and Cowley, A.D., Barometric oscillations, In Introduction to Seismology (ed. Bath. M.) (Birkhäuser, Basel 1973) pp. 193-222.

Submission of Manuscripts

Manuscripts must be submitted in triplicate, formatted with double line spacing and wide margins. Copies of the figure should be attached at the end of the manuscript. Original, high-quality, glossy figures may be submitted later. All manuscripts pages, including references, tables, and captions, should be numbered consecutively, starting with the title page as page one.

All manuscripts should be submitted to Regular Issues Editor-in-Chief
Brian Mitchell
Department of Earth & Atmospheric Sciences
Saint Louis University
3507 Laclede Avenue
St. Louis, MO 63103, USA
e-mail: mitchell@eas.slu.edu

Delivering manuscripts in diskette form may substantially facilitate the publication process provided certain points are taken into consideration:

Texts on diskette should be delivered in either DOS or Macintosh format. They should be saved and delivered in two separate versions:
• with standard text format as offered by your word processing program, and
• in addition in Rich Text Format (RTF) or, as a last resort, as an ASCII file.
Numerous word processing programs offer these options when saving the text.
The final hard copy of the manuscript should be submitted together with the diskette.
The electronic and printed version must be absolutely identical. All pictorial and graphic illustrations should be delivered as hard copy originals and must be 200% of the final printed size. Digital drawings and graphs should be submitted in Encapsulated PostScript (EPS) or Tag Image File Format (TIFF) form. Do not fail to include a hard copy for ready viewing. Back-up copies of the diskettes must be kept. Diskettes must be adequately protected for transport.

Galley Proofs
Unless indicated otherwise, galley proofs will be sent to the first-named author directly from Birkhäuser Verlag AG and should be returned with the least possible delay. Textual alterations made in the galley proof stage will be charged to the author. One copy of the corrected proof is to be returned immediately to
Editorial Office
Attn. Mrs. Renate D'Arcangelo
Harvard University
233 Pierce Hall, 29 Oxford Street
Cambridge, MA 02138, USA

The editorial office assumes no responsibility for delayed proofs, errors in the original manuscript, or major alterations in proofs for any reason.

Reprints
The authors will receive 50 reprints of each article without charge. Additional reprints may be ordered in lots of 50 when the final corrected page proofs are returned. Orders submitted thereafter are subject to considerably higher rates.